Perspectives on the Popularisation of Natural Sciences in a Diachronic Overview

Perspectives on the Popularisation of Natural Sciences in a Diachronic Overview

Edited by

Eleonora Chiavetta and Silvana Sciarrino

CAMBRIDGE SCHOLARS PUBLISHING

Perspectives on the Popularisation of Natural Sciences in a Diachronic Overview,
Edited by Eleonora Chiavetta and Silvana Sciarrino

This book first published 2014

Cambridge Scholars Publishing

12 Back Chapman Street, Newcastle upon Tyne, NE6 2XX, UK

British Library Cataloguing in Publication Data
A catalogue record for this book is available from the British Library

ISBN (10): 1-4438-5498-0, ISBN (13): 978-1-4438-5498-6

TABLE OF CONTENTS

LIST OF ILLUSTRATIONS

LIST OF TABLES

PREFACE

This book brings together carefully selected articles written by scholars in the field focusing on specialized knowledge. The volume presents some of the relevant findings of a National Research Project that has involved researchers from several Italian universities. The main focus of the project was to analyse discursive popularisation mainly in the context and domain of natural sciences viewed in a diachronic perspective. Scholars involved in the project have focused their studies on the creative transformation, hybridisation, and even bending of genres used to popularise scientific discourse for different communicative purposes and audiences, thus extending the conventional genre boundaries to disseminate specialized knowledge.

The book combines strands of research which are currently debated in linguistic scholarship –the issue of specialized discourse, the issue of knowledge dissemination and the issue of the versatility of genres. The book includes six in-depth studies focusing on a wide range of fields and genres.

Daniela Cesiri investigates the popularisation of terminology in Irish botany texts published during the Late Modern English (LModE) period. She analyses how the opening of botany as a scientific discipline to a public of amateurs and to their contribution in collecting and classifying samples, plays a very relevant role in shaping the popular botanical terminology. Eleonora Chiavetta also focuses on botany, and analyses the features of a very popular genre of garden literature, namely the garden notebook. The main feature of this genre is the fact that a gardening and botanical knowledge is popularised in a chatty, informal way, which combines scientific and professional competence and expertise, with personal comments and references to the author's private sphere. Two major examples of the genre are considered: the successful series of books by Mrs. C. W. Earle-*Pot-Pourri from a Surrey Garden* (1897), *More Pot-Pourri* (1899) and *A Third Pot-Pourri* (1903)-and *Country Notes* (*1939)* by Vita Sackville-West. Thomas Christiansen carries out an analysis of how Darwin's theory of evolution is presented or misrepresented in George William Hunter's famous high school textbook, *A Civic Biology Presented in Problems*(1914). Christiansen's corpus analysis concentrates on speaker stance, including the interpersonal functions, and particularly on the way in which ethos plays a major role in the presentation of

Darwin's theories. Gabriella Di Martino focuses her research on the plain use of scientific language in the 17th century. She compares different versions of the work by Joseph Glanvill (1636-1680), the elaborate essay *The Vanity of Dogmatizing*. When this work was attacked by the Aristotelians, Glanvill answered by issuing a revised version, *Scepsis scientifica*, which defended the experimental method of the Royal Society and argued for a plain use of language. Kim Grego's study is based on a corpus of British newspaper articles (1880s-2000s) dealing with Down House, Darwin's museum-house. The vicissitudes undergone by this house are used to navigate through the alternating fortunes of Darwin's theories, ideas and works in the 20th and 21st century. In this way, a novel perspective is given from which to look at the popularisation of science. Finally, Stefania Maci's research aims at disclosing in what manner and to what extent scientific knowledge is popularised by members of the medical academic community for an academic audience. With this objective, Maci analyses two Nobel Prize lectures held in 2009, by Elizabeth H. Blackburn and by Carol W. Greider, and compares them with the articles the two Nobel winners wrote, in which their discovery was scientifically described.

The editors wish to thank all the contributors of the present volume. Heartfelt thanks goes to Ms Hilary Caminer (The Open University), for her precious suggestions and supportive advice.

CHAPTER ONE

POPULAR BOTANICAL TERMINOLOGY IN IRELAND DURING THE LATE MODERN ENGLISH PERIOD: A DIACHRONIC OVERVIEW

DANIELA CESIRI

1. Introduction

The present investigation deals with popular botanical terminology in Ireland during the Late Modern English (LModE, 1720 c.-1910 c.) period. The main project[1] behind this research which has been conducted in three phases started in 2011 with the aim of studying a specific genre, i.e.botany texts written by specialists but addressed to amateurs.

The particular focus on Ireland derives from the fact that so far no linguistic study has taken into account a diachronic overview of the evolution of popular botanical terminology in texts published in Ireland. Findings from the project could constitute further contribution to the already available literature on the terminology used for the local flora in the "British Isles" (e.g., Elliston and Hatfield 2004) and in overseas countries (e.g., Masiola and Tomei 2009).

In addition, the LModE period is of particular interest because it saw the establishment of botany as a scientific discipline and also because botany became a popular activity for the middle classes who collected samples and classified them using popular terminology in English (cf. Constantine 1981).

[1]This is part of a main project entitled "The Popularization of Scientific Discourse in a Diachronic Perspective", co-financed by the Italian Ministry of Education, University and Research (prot. n. 2008C7BR9H_002).

The present chapter is the third and last phase of the project and seeks to present popular botanical terminology in English as it was used at the end of the nineteenth century in Ireland as well as to compare findings from the two previous phases, published in Cesiri (2012a) and Cesiri (2013b), in order to present a diachronic overview regarding the whole period under consideration.

2. Previous Findings

Each of the three phases of the project considers a sub-period of the general LModE period in order to investigate better the evolution of the popular botanical terminology. The first phase considered the second half of the eighteenth century (henceforth "Early LModE"), the second phase considered the first half and the middle of the nineteenth century (henceforth "Mid-LModE"), while the present third, and final, phase considers texts published at the end of the nineteenth century ("Late LModE").

The collection of texts pertaining to the sub-periods being taken into consideration was quite problematic since, unlike publications regarding England or other English-speaking areas, it was quite difficult to find the type of works required for the projects' purposes. Indeed, as already explained in Cesiri (2012a) and Cesiri (2013b), the books needed for the study had to be volumes on botany written by specialists but for a public of amateurs whose interest focused on the semi-professional technical nature of the discipline, without any hint to the popular folklore and legends attached to the plants themselves. In addition, the texts had to deal with species found in Ireland to differentiate their terminology with that inserted in other publications generally presenting species found in the "British Isles". These publications were not considered to provide genuine information regarding popular botanical terminology in Ireland, also in consideration of the particular linguistic situation of Ireland with English, Latin and Irish Gaelic all available to name the several species (cf. Cesiri 2012a).

The specific nature of the texts required for the study and the difficulty in tracing publications inaccessible to the public, either because they were not preserved and lost through the passing of time or because held in private collections, account for the limited number of texts traceable. Indeed, the first phase could investigate only one text for the Early LModE period (K'Eogh 1735) and three texts for the Mid-LModE period, namely Mackay (1836), Power (1845) and Dickie (1864).

For a more detailed account of the selection process for the texts, the reader can refer to the two earlier publications (Cesiri 2012a and 2013b) as this section is principally meant to summarise the findings from the first two phases which will be particularly linked to the present phase later on in the chapter. In particular, Cesiri (2012a) found that the Early LModE text employed a pre-Linnaean system of classification based on the work of other botanists or natural theologians. K'Eogh presents his readers with a tripartite terminology, in which Latin, English and Irish Gaelic are used to indicate the name of each species and genus. In addition, the terminology of the 1735 text in English shows an extensive use of what Gotti (2003: 73) defines as "nominal adjectivation", consisting of nouns in adjectival function that specify other nouns, a use which is also very frequent in present-day scientific texts (see Hughes 1988, cit. in Gotti 2003: 74).

In the second phase of the project, Cesiri (2013b) discovered that the three Mid-LModE texts, employing a Linnaean system of classification, show the same recurrent use of nominal adjectivation, thus confirming that this phenomenon of word-formation might have become a pattern typical of that specific domain. In particular, in K'Eogh (1735) the majority of entries are constituted of one-word terms (e.g., "agrimony", Lat. *Agrimonia*), including two terms linked by the hyphen but reported as one single word (e.g., "dog-berry tree", Lat. *Ribes Cynosbati*), but the text also shows a frequent use of two-words terms (e.g., "manur'd flax", Lat. *Linum Sativum*); whereas the three Mid-LModE terms show the tendency to use predominantly single terms (e.g., "Cowslip", Lat. *Primula Veris Major*), which is probably to be interpreted as an increased preference for conciseness.

In addition, the texts of this period show a process of standardisation or even of dialect levelling in their use of English and Latin only, whereas Irish Gaelic is completely absent. A final difference between the Early LModE text and the Mid-LModE texts is their textual structure as the former also lists the remedies for which a plant or herb can be used, whereas the latter present only the plant, its physical aspect and geographical distribution. A greater uniformity is to be found in the three texts of the Mid-LModE period, thus probably showing a standardisation also in the structure of the texts as specific genre since their nature seems to develop from a popular herbal into more scientifically-oriented and professionally-organised texts (see Cesiri 2013b).

The investigation of the texts to be found for the Late LModE period, then, will seek to complete the picture presented so far in the first two phases. As already mentioned in the Introduction, the present chapter will first describe the texts found for the end of the nineteenth century and, then, will seek to complete the diachronic overview of the evolution in popular botanical terminology in Ireland.

3. Amateur Botany and Botanists at the End of the 19th Century

The phenomenon of amateur botany became increasingly popular during the nineteenth century in England as in Ireland (cf. Constantine 1981). If the previous century saw an interest in the medical and culinary properties of herbs, the following one experienced a changed interest towards a more scientifically-oriented approach to plants collected for their beauty but also as scientific samples of local fauna. These collectors were generally men, but women were also involved; they belonged to the middle and upper middle classes and, having no field training or academic knowledge, they relied on the works of experts to guide them in their collection (cf. Constantine 1981). As the Mid-LModE and the Late LModE texts openly admit, the experts were sometimes helped by the same amateur botanists who collected unknown species and had them classified by the experts who gave them a Latin name and inserted the new sample in their books, giving open and full credit to the amateurs for the discovery and the popular name in English. This virtuous circle contributed to the dissemination of specialized knowledge through popularising works as it actively involved both specialists and amateurs in the creation and dissemination of a technical terminology for the general public (cf. Synnot 1997).

As far as the Late LModE period is particularly concerned, Praeger (1934: i) defines Ireland as “a pleasant country for the botanist” because it contained fewer varieties than England but of greater beauty and greater specificity in the species. In addition, the discovery of new, native species by the Royal Dublin Society increased the popular and professional interest in Irish botany (cf. Webb 1986). This followed a new cultural fashion of the time since non-experts with an average education participated in public debates about Darwinism, a signal that the natural sciences were opening to the public (cf. Mordavsky Caleb 2007). This ultimately extended to other disciplines, such as archaeology (see Cesiri 2012b) and botany itself. As a consequence, the period saw the publication of a greater number of books addressed to amateurs, who were actively involved in fieldwork and, thus, contributed greatly to the discovery of new species and to the creation of popular terms which were, later, given their Latinate equivalent by professional botanists (cf. Synnott 1997).

4. The Present Study: The Texts under Investigation

The texts analysed in the present chapter share, with the other texts analysed in the rest of the project, the nature of texts written by specialists to amateurs. The texts are Stewart and Corry (1888; 193, 776 words) and Chichester Hart (1898; 92, 288 words), both reportedly(see, for instance, Webb 1986 and Synnott 1997) considered significant contributions to Irish amateur botany of the Late LMod period.

Analysis of their textual organisation indicates that they show a common structure, starting with a Preface which contains thanking words and acknowledgements of the work of other experts and amateurs who helped in the collection of some samples or in the discovery of new species. An Introduction follows, occupying one third of the whole volume and containing a thorough description of the history, geology, topography and climate of Ireland (or a specific part of its territory). The following section is called, in both texts, "Sources and Contributors" with a list of names of the helpers mentioned in the Preface. The "List of Genera and Species" is finally introduced without providing any indication about the system of classification used (unlike the Mid-LModE texts), probably because all the authors follow the well-established Linnaean system, by this time considered the system of classification generally acknowledged as standard by the scientific community of botanists (cf. Stuessy 2009).

Some further remarks should be made on the actual textual structure of the texts, concerning particularly the Introduction, in which the addition of climate, geology, topography and so forth (absent in the Mid-LModE texts) provides a more professional organisation and outlook to the Late LModE volumes. At the same time it is clear that it is meant for amateurs as the language is simplified and serves to provide the amateur reader with the tools for recognising particular soils and environments in which similar species usually grow to differentiate one species from similar ones, through an exclusion process. Hence, if a sample is found growing in an unusual soil or in a climate other than the norm, the amateurs are able to recognise it as a new discovery or – more frequently – as a plant belonging to a particular species or genus but typical of that area or climate.

Finally, the volumes show a very limited use of drawings or schemes representing the main genera of the plants and their parts. They are inserted only in the initial sections of the volumes, unlike the Early LMod and Mid-LMod texts, which contain an extensive use of drawings in all their sections. This might be due to two different reasons: one might be called "editorial" as an extensive use of pictures might have increased the production costs and the final cost of the volumes, making them affordable only for an elite of readers losing their nature of "popular" publications for the middle classes; the second reason might be called "didactic" as the authors might want to provide their readers only with schematic visual representations of the genera, leaving to them the task of recognising the species from their own field experience as well as preferring to describe in words the aspect and composition of the plants. However, no explanation for this absence of figures is provided by the authors or in the relevant literature, hence the absence of pictures could be the norm of the time for the genre and the interpretation here provided can be considered only tentative.

4. 1. Textual Structure of the Late LModE Texts

Figures 1.1.and 1.2. show sample pages from the two Late LModE texts under investigation. As the structure in which the entries are organised and presented is fundamental to this kind of text (meant to be manuals for easy consultation), a comparison between the Late LModE texts and the texts from the two previous phases will be provided in order to detect any evolution or modification in the textual organisation of the entries themselves.

First of all, we can notice that the division of the content into paragraphs is organised thematically according to the places in which the species can be found and to the type of soil in which it grows. The authors, especially Stewart and Corry (1888), also insert some reference to the place where the species was first recorded, whereas Chichester Hart (1898) adds some more precise dating; in addition, the latter seems to rely more on previous volumes as well as on previous other collectors than Stewart and Corry's volume that quotes mainly the two authors' "discoveries".

If we compare the Late LModE texts to the Early LModE and to the Mid-LModE texts (shown in Figures1.3. and 1.4.), the main differences we can notice between the three different sub-periods lie not only at the level of text structure but also at the level of content distribution and authors' aims. Mackay (1836), representative of the three Mid-LModE texts, presents itself as one single block, whereas the Early LModE and the Late LModE texts present a division functional to the purposes of the authors, practical in both cases but meant for easier consultation of the herbs' usage in medicine or cooking (in K'Eogh 1735) and for easier recognition of their different characteristics (in Stewart and Corry 1888 and in Chichester Hart 1898).

FLORA OF THE NORTH-EAST OF IRELAND.

ORDER I. RANUNCULACEÆ.

THALICTRUM *Linn.*

1. T. minus *Linn.*

On blown sands by the seacoast; usually abundant where the conditions exist. Fl. June and July.

Down—Sandhills near Newcastle; *Templeton*, 1793; and T.H.C., 1878. In profusion on Ballykinler sands east of Dundrum; S.A.S.

Antrim—Sparingly at Ballycastle; T.H.C. Portrush; S.A.S.

It is remarkable that this plant has not been found at Magilligan, Co. Derry, the wide expanse of sandhills being so eminently favourable.

Var. β T. MONTANUM *Wallr.*

Damp shady rocks—very rare. Fl. July and August.

Down—Mountains of Mourne, plentifully; *Harris's Down*—1744. West side of Bignian Mountain; *Templeton*, 1808. Rocks above Donard Lodge, at 1,500 to 1,600 feet; S.A.S. Kilbroney River north of Rostrevor; R.Ll.P.

Antrim—On basaltic rocks in Glenariffe; *Cyb. Hib.* The Portmore, and Lough Beg plants of Flor. Ulst. were *T. flavum.*

Derry—Artrea near Toome; D.M.

2. T. flavum *Linn.* MEADOW RUE.

Marshy ground, and wet meadows—rare; about Lough Neagh, only. Fl. July and August.

Antrim—Portmore; *Templeton*, 1800. Massereene Demesne; *Flor. Ulst.* Ram's Island; *F. Whitla.* Shane's Castle; T.H.C. Cranfield Point; S.A.S. Selshan Harbour; *G. C. Hyndman.*

Derry—Artrea near Toome; D.M.

Figure 1.1. Page sample from Stewart and Corry (1888).

PHANEROGAMIA.

DICOTYLEDONES.

Ranunculaceæ.

Thalictrum *Linn.* MEADOW-RUE.

1. **T. alpinum** *Linn.*

Wet alpine rocks and by spongy rills. At a considerable height on mountains in the west and south west. Rare. Fl. July, August.

IV. Poisoned Glen, at 1,200 to 910 feet; in a gully on Slieve Snacht Beg at 1,800 feet, (see under Oxyria reniformis); Breaghy, above Lough Altan, at about 900 feet.

VIII. At 1,900 feet a mile west of Lough Belshade; north and south sides of Lavagh[illegible], and on Silver Hill, 2,150 to 1,400 feet in the Bluestack Mountains; Slieve League, 1,550 to 1,310 feet; at 1,400 feet on Alt Mountain, south of Ardara; Croaghnaclerragh at 1,200 feet, south of Ardara; Croaghorm, near Martin's Bridge, at 900 feet and above. From 2,150 to 900 feet above sea level. Forming the sod on Croaghorm in many places.

First record; Hart, 1879. (*Jour. of Bot.*)

2. **T. minus** *Linn.*

Broken limestone, rocky ground, in one small area. Very rare. Fl. June to August.

VII. Abundant on broken limestone ground south of Brown Hall, near Ballintra. Typical.

First record; Hart (*Flora*).

T. dunense Dum.

Sandhills and thickets on sandy soil near the sea. Very rare.

IV. Sandhills at the mouth of the Ray River, near Falcarragh, an unusually small, condensed form, with the florets of the normal size. I have not seen anything like it in Ireland before. This is the only locality in Donegal for the sandhill Thalictrum. July, 1894.

First record; Hart (*Flora*).

Figure 1.2. Page sample from Chichester Hart (1898).

In the Late LModE texts (as in the Mid-LModE texts) the classification of terms has a purely "scientific" purpose whereas in the Early LModE text the author shows also a linguistic-etymological interest especially in the Irish Gaelic language, completely absent in the later texts, which show a progressive standardisation of terminology.

> ORD. 42. BORAGINEÆ. *Juss.* Borage Family.
>
> Calyx with 5 (rarely 4) divisions, persistent. Corolla hypogynous, monopetalous, mostly regular, 5, sometimes 4-cleft; its æstivation imbricated. Stamens inserted upon the corolla, of the same number with its segments, rarely more. Ovary 4-partite, 4-seeded or simple, and 2—4-celled; ovules definite, pendulous. Nucules 4, distinct or combined below; sometimes a 4-celled drupe, or a berry with 2—4 bony seeds. Seeds without albumen or nearly so. Embryo inverted.—*Herbs or shrubs. Leaves alternate, without stipules, generally rough. Flowers frequently in secund spikes, sometimes panicled or corymbose, sometimes axillary, solitary.*

Figure 1.3. Sample entry from Mackay (1836).

> ftructions.
>
> BORRAGE, Hib. *Boraflegorrum.* Lat. Borrago, it grows in Gardens, and Flowers in *June.* It Comforts the Heart, makes Men Chearful, and Merry; therefore drives away Sadnefs and Melancholy, which agrees with this Old Verfe.
>
> *Cordial*
>
> *Ego Borrago, Gaudia, Semper ago.*
>
> *One of the four Cordial Flowers.* It is *Alexipharmic,* good againft *Malignant Fevers, Small-pox, Meafles,* &c. Boiled in Hydromel, is good for a *Raucidity* or Hoarfnefs.
>
> BOX-TREE Lat. Buxus. The Leaves of Box are hot dry

Figure 1.4. Sample entry from K'Eogh (1735).

Considering the discourse organisation of the entries, we can notice a narrative with full sentences in K'Eogh (1735) and an almost telegraphic writing style in Mackay (1836), whereas the Late LModE texts contain both these features as they describe the content in full sentences but also include passages in a telegraphic writing style especially when they describe the appearance of the plant.

The texts in all the three sub-periods make use of a specific terminology, but not a fully technical one, along with recurrence to lexis from general language, typical of publications presenting specific contents to a public of amateurs. A further distinction can be found comparing K'Eogh (1735) with the texts of the later sub-periods. The Early LModE text shows, indeed, a description of plants aimed at their employment in domestic usage (medical remedies or cooking recipes), while the later texts all show a clear popularising purpose of scientific content.

5. Research Rationale and Results

As already mentioned, this chapter aims at analysing the popular botanical terms in English as used in the Late LModE texts, looking for patterns of word-formation and comparing these results to the ones obtained in the previous phases, thus completing the diachronic overview of Irish texts published during the LModE period.

The two volumes under investigation were searched manually for a precise categorisation in the composition of the entries. This count provides an overall picture of the most recurrent word-formation patterns and is completed by a qualitative analysis of the categories thus found.

5. 1. Quantitative Analysis

Table 1.1.and Table 1.2 . below illustrate the patterns in word-formation found in the two Late LModE texts in raw figures and in figures normalised to 10, 000 words, respectively. Normalised figures are provided to give a more accurate and representative frequency of occurrence in comparing the two texts, their different word count considered. The terms were divided into two separate groups. The first, labelled "Genus/Main Species", contains terms used to refer to the main genus or species of a plant, whereas the "Sub-Species" group contains terms for the sub-species of the same plant reported for some samples but not for all.

The different terms were classified according to their word-formation pattern, hence "1 word" refers to single-word terms (e.g., "Sundew", Lat. *Drosera*), whereas "1 word (-)" indicates two terms linked by the hyphen but considered as one single term by the authors themselves (e.g., "Skull-cap", Lat. *Scutellaria*). The label "2 words ('s)" includes two terms linked by the "*s*-genitive" (cf. Biber *et. al.* 1999: 300ff; e.g., "Shepherd's Needle", Lat. *Scandix*).

The other categories include: "2 words (1+2-)", i.e.two terms of which one is composed by a hyphenated compound of two words (e.g., "Mouse-ear Chickweed", Lat. *Oerastium*), "2 diff. words" or two separate terms (e.g., "Globe Flower", Lat. *Troilus*); "3 words" is the category including three terms forming the name of a plant (e.g., "Dwarf Elder Danewort", Lat. *Sambucus Ebulus*). Finally, "Phrase" includes terms composed of two or three words, composing a phrasal-like term (e.g., "Robin-run-the-hedge", Lat. *Gallum Aparine*, or e.g., "Grass of Parnassus", Lat. *Parnassia*).

In addition, some figures in Table 1.2 were highlighted to stress the most relevant patterns occurring in the two Late LModE terms;namely, figures in bold type indicate a significant frequency, figures in underlined bold type indicate patterns with a lower frequency and, finally, in italicised bold type are indicated the figures with a potentially "borderline" relevance as they present a non-significant frequency in one text with respect to the other one.

Text	Genus/Main Species							Sub-Species				
	1 word	*1 word (-)*	*2 words ('s)*	*2 words (1+2 -)*	*2 diff. words*	*3 words*	*Phrase*	*1 word*	*1 word (-)*	*2 words ('s)*	*2 diff. words*	*Phrase*
Stewart & Corry (1888)	342	5	20	--	64	1	5	--	--	4	101	--
Chichester Hart (1898)	166	28	15	3	65	2	3	41	13	11	79	4

Table 1.1. Raw figures

Text	Genus/Main Species							Sub-Species				
Norm. per 10,000	*1 word*	*1 word (-)*	*2 words ('s)*	*2 words (1+2 -)*	*2 diff. words*	*3 words*	*Phrase*	*1 word*	*1 word (-)*	*2 words ('s)*	*2 diff. words*	*Phrase*
Stewart & Corry (1888)	**17.65**	0.26	***1.03***	--	**3.30**	0.05	0.26	--	--	0.21	**5.21**	--
Chichester Hart (1898)	**17.99**	**3.03**	***1.63***	0.33	**7.04**	0.22	0.33	**4.44**	***1.41***	***1.19***	**8.56**	0.43

Table 1.2. Figures normalised to 10, 000.

As can be seen from the tables, Stewart and Corry (1888) contains a greater number of terms constituted of single words for the "genus/main species", followed by terms of two different words for "sub-species". Chichester Hart (1898), on the other hand, reports a greater variety of terms with a distinct frequency for terms of one word, followed by terms of two different words, hyphenated compounds considered as one single term, and a minor presence of "*s*-genitives". As for "sub-species", in the 1898 text we have a more frequent occurrence of terms composed of two different words and one-word terms, followed by a minor occurrence of hyphenated compounds considered single terms and "*s*-genitives".

The distribution of terms in the two Late LModE texts can be represented visually inTable 1.3. and Table1.4. Table 1. 4. in particular, represents the normalised figures and offers the reader greater precision in visualising the distribution of word-formation patterns, with single-word terms clearly outnumbering other patterns in both texts.

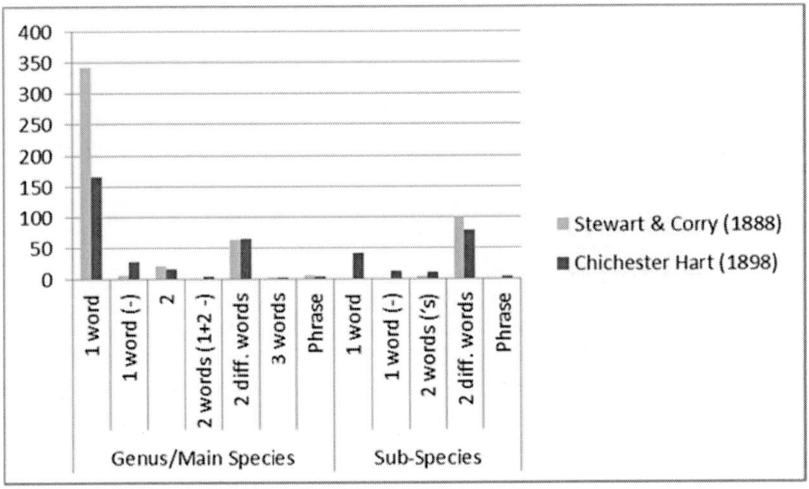

Table 1.3. Distribution of raw figures.

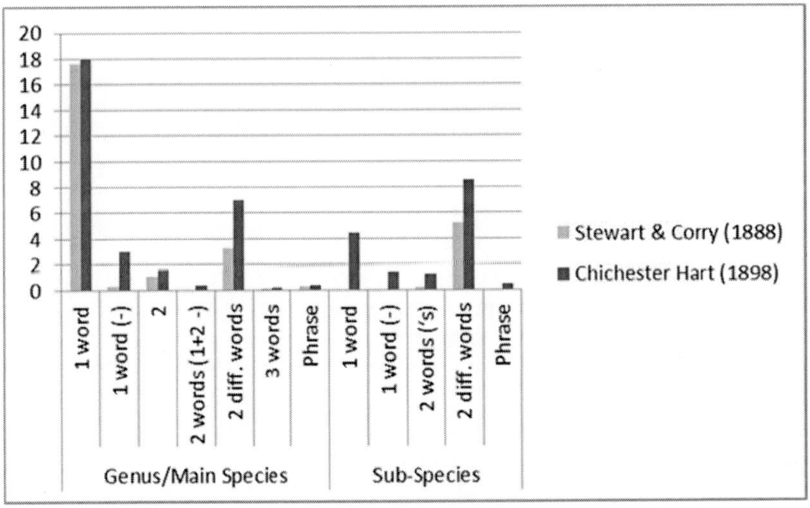

Table 1.4. Distribution of figures normalised to 10, 000.

5.2. Qualitative Analysis

This section provides examples of the different categories in the terminology of the two Late LModE texts. This qualitative analysis will focus on the word-formation patterns only, as the discourse structure and the narrative of the paragraphs explaining the plants' characteristics is not of interest to the present analysis. It might, however, constitute an interesting object of study for a future analysis considering the syntactic and stylistic evolution of this genre within the broader context of specialized discourse.

As already seen in the qualitative part of the analysis, one-word terms are the most frequent category occurring in the two texts. Examples (1) to (3) show the different sub-categories in which the single-word terms can be classified; they also show that this type of term often includes compound words (examples 1 to 3) which contain the union of two nouns (N+N) or an adjective and a noun (Adj+N), describing the aspect, the qualities or the properties of the plant. Animal-based terms are particularly frequent, as in the list in example (4).

(1) "Simple" terms: Agrimony, Barley, Daisy, Ivy, Holly.
(2) "Descriptive" terms: Blackthorn, Mayflower, Pondweed, Watercress, Windflower.
(3) "Creative" terms: Goldilocks, Danesblood, Hairgrass, Selfheal, Eyebright.
(4) "Animal-based" terms: Cowberry, Crowfoot, Hawksbeard, Hogweed, Horsetail, Toadflax, Wakerobin.

A greater use of hyphenated compounds as single words is made in Chichester Hart (1898) than in Stewart and Corry (1888), but this might be explained by his relying more on previous texts (such as Mackay 1836 and Dickie 1864) than on amateurs, the author's own fieldwork and the general public for popular terms.

Terms in examples (5) to (8) are divided according to word-formation patterns and represent the other categories found in the terminology of the Late LModE texts. Examples (5) to (7), in particular, show that the main noun in a term is usually pre-modified by either an adjective or a noun which serve to provide an almost visual representation of the variety, the characteristics or the general aspect of the plant but they might also refer to the animals or to the popular folklore connected to a particular plant. Terms in example (8) illustrate instances of those which show a more complex, phrasal-like structure than the terms in other categories. These terms also represent some extremely creative and "iconic" representation in the word-formation processes of popular botanical terminology in the texts.

(5) "2 diff. words": Bastard Pimpernel, Fairy Thimble, Globe Flower, Tormenting Root.
(6) "1 word (-)": Cow-Wheat, Mat-Grass, Moor-Grass, Skull-Cup.
(7) "2 words ('s)": Cat's Tail, Viper's Bugloss, Sheep's Sorrel, St. John's Wort, Hare's Foot Trefoil, Our Lady's Bedstraw, Shepherd's Needle, Enchanter's Nightshade, Farmer's Plague.
(8) "2 words (1+2)": Ivy-leaved Crowfoot.
(9) "Phrases": Robin-run-the-hedge, Jack-by-the-hedge, Forget-me-not, Flower of Dunluce, Grass of Parnassus (sub-species of Poor man's blanket).

6. A Diachronic Comparative Overview (From 1735 to 1898)

The final phase of the present chapter, as well as of the research project in general, provides an overview of the diachronic evolution of the word-formation patterns in popular botanical terms in Irish texts of the LModE period. It will be conducted by comparing results from the three sub-periods taken into consideration, namely the Early, the Mid-and the Late LModE periods, respectively.

The first "evolution" to be found in the corpus of texts investigated involves the level of their textual structure since they show a shift from the popular genre of herbals to the genre of semi-professional monographic volumes, a change brought by the different public for whom the volumes were meant. This new public was usually more educated than the one for which the 1735 text was meant and was trained to collect field samples. The public of amateurs actually demanded the latter kind of volumes for their own personal development in scientific subjects as they presented scientific reliability in a popularised form, which they could follow when collecting their own samples in the fields and woods.

As for the actual nomenclature found in the texts, considering the category of one-word terms, we can affirm that terminology remained substantially unchanged since the 1735 text but it does show a change in its frequency of occurrence, the later texts showing an increasingly greater frequency in the use of single-word terms.

Two-words terms keep their patterns over time, that is to say that the N+N and the Adj+N structures remain the preferred ones as they serve to identify concisely, but effectively, the qualities, properties and features of the plants or the soil on which they grow, the relation to the fauna or to the local folklore. Their frequency is again the factor which changes over time as their number gradually decreases in the Late LMod texts probably because they "evolve" into single terms, especially in the form of hyphenated compounds.

Finally, a further diachronic change involves the number of alternative terms, or synonyms, provided for a species. In particular, the number of synonyms reported for a species decreases over time, hence probably testifying to a reduction in terms of lexical richness, i.e .a "lexical reduction" in the variety of plants' popular names. In turn, this reduction might be explained as a tendency towards conciseness as well as monoreferentiality in the terminology that signals a preference for these particular lexical features typical of specialized language (cf. Gotti 2003), though in the popularised form of scientific botanical language.

7. Conclusions

To generalise the implications of the findings from the research, we might underline that changes occurring in the popularising language of botany in the LModE period parallel, and are paralleled by, changes occurring in the nature of the discipline itself during the same period. As explained at the beginning of the present chapter, the opening of botany as a scientific discipline to a public of amateurs in general, and to their contribution in collecting and classifying samples in particular, plays a fundamental role in shaping the popular botanical terminology. This is because the terms in English start to be taken from popular folklore by the amateurs or they are coined anew by the amateurs themselves on the basis of the function, place or features of the samples collected. At the same time, though, the amateurs have access to the scientific, specialized aspects of the discipline, hence they learn how to term new samples in a more professional and less intuitive way. This, however, does not diminish the creative force of some of these terms.

The increasing preference for a single-term terminology and reduction in synonyms implies also a preference for the features of conciseness and monoreferentiality typical of scientific discourse. In this nomenclature, terms not only acquire a more professional and rigorous aspect but they also become easier to remember as well as to associate to a particular plant; at the same time the species become easier to recognise and name during field work by "new generations" of amateur botanists.

The present study has also shown that along with nominal adjectivation, which was already detected as typical of the Early LModE text, some other lexical features characterising present-day specialized discourse emerge since the "beginnings" of the discipline, even though we deal with popular terminology by and for amateurs.

Moreover, we can mention a "return" to local terms in the Late LModE texts but the authors firmly keep them in an extremely marginal position, too marginal and secondary to be quantified and categorised especially with respect to the rest of the terminology which is in a general, "standard" popular language. Indeed, the authors themselves do not indicate whether the terms are from local Irish English dialects or from dialects of English. The present author herself, after consultation of reference sources such as the OED, Wright (1898-1905) or Dolan (1998), could not ascertain the nature and origin of this very limited group of synonyms. In addition, the absence of terms in Irish Gaelic in texts from both the Mid-LMod and the Late LMod periods is also significant in that it can imply that the Gaelic revival, taking place during the two sub-periods, was propelled by the cultural and literary circles of Ireland[22] but it did not involve the community of Irish amateur or professional botanists.

This can be interpreted in the light of another study conducted in Cesiri (2013a) and comparing the community of Irish amateur botanists in Ireland to that of English botanists in England during the LModE period. Findings from the study revealed that more important than local terminology was the adoption of the norms connected to and accepted by the "scientific" community of botanists and amateurs regardless of the country where they "operated". Hence, the absence of terms in Irish Gaelic in the Mid-LMod and Late ModE texts appears to be in line with this general preference for adhering to the norms of one's specific, "supra-national" scientific or professional community rather than to the cultural and social norms of one's country-specific, culturally-bound local community.

[2] The so-called "Gaelic Revival", or "Gaelic Revivalism", is defined by scholars as a cultural and literary phenomenon, whose "political relevance … is less easy to establish" (Foster 1989: 446). It is, indeed, considered a continuation of "the long tradition of Celtic antiquarianism" (Ibid.) which was adopted by the nationalist movement at the end of the nineteenth century to provide its claims to Irish national identity and sovereignty the necessary connection to the roots of the ancient Gaelic society to which patriots were constantly referring in their fight for independence from the English crown. In addition, the "Romantic" re-mediation of Gaelic myths and Celtic folklore (undergone through literature but also through the rediscovery of the Celtic language) provided modern Ireland with a "consciousness of itself" (Kiberd 2001: 407), of its past, "almost disappeared in the decade of the Great Famine" (Ibid.).

In conclusion, the study conducted and the project completed in the present chapter have shown that in the LModE period, as in the present day, the language of a specific domain is influenced by changes in the community to which it refers, as well as by changes in the related discipline, but it appears to be relatively isolated (with particular emphasis to be put on the adverb) from changes of a different nature occurring in the country where members of this community, either professionals or amateurs, are active. This appears to be valid not only for the specialized language characterising the scientific domain of botany but also for its popularised, and popularising, language.

CHAPTER TWO

PRIVATE AND NON-PRIVATE FEATURES IN GARDEN NOTES AND NOTEBOOKS: MRS. C. W. EARLE AND VITA SACKVILLE-WEST

ELEONORA CHIAVETTA

1. Introductory Remarks

There is a great variety of genres in British gardening literature. There are scientific and scholarly texts on horticulture and botany, as well as practical manuals on gardening matters and books on the history of gardening or on historical gardens. Various innovative genres have been introduced influenced by the cultural and social climate of the time and have become successful. Examples are botanical dialogues, didactic books written as a collection of letters, dictionaries, encyclopaedias. Hoyles (1994: 55) divides popular gardening writing into two main areas, and distinguishes discursive books from handbooks whose aim is to inform and instruct. Gardening handbooks and manuals have been written to satisfy the users' needs since the Middle Ages. The earliest known is the manuscript by"Mayster Ion Gardener" probably compiled near the middle of the fourteenth century by a master gardener working at Windsor Castle or at the Royal Palace of Westminster (Harvey 1985: 92). The author was an expert and the objectives of the text, which was probably meant as a memorandum, were practical-there are, for example, instructions on the grafting of trees and on the sowing of seeds (Hadfield 1960: 30).

The first genre of historical interest is probably the herbal-whose aim was to catalogue English plants-which was studied in England as early as the ninth century. The oldest surviving example is the Saxon work, Bald's Leechbook, dating from about A. D. 900-950 (Rhode 1922: 1-5). Although herbals were "not strictly gardening books", as Martin Hoyles observes, since they mainly focused on the medical properties of plants, "they also had a close connection with gardening as the authors usually had gardens of their own" (1994: 55). Today, they are extremely useful sources for data about what was actually planted in gardens in the various periods (Uglow 2005: 75). Herbals are, in fact, a rather stable genre, as they have been written and published in every century up to the present day, and have given origin to other text types: Susan Kermas (2006), for example, sees a connection between herbals and advertising texts.

One of the commonest discursive genres used by gardening writers is the garden notebook, that is a book in which the authors, who may be either experts or amateurs, mainly write notes about things they have to do or have already done in the garden, but also tell the reader about their own gardens, describing changes and development over time. Since the main focus of such texts is often the life-story of the garden, this text type has been considered "a garden autobiography" (Seaton 1979). At the same time, while describing their garden, the authors also reveal something of their personality, by expressing their opinions on gardening matters, on nature, and often on life in general.

Beyond the descriptive aims of garden notebooks, an important objective of this popular non-fiction is to convey specialist knowledge with informative and educational purposes by offering advice and instruction to a wide amateur readership. What distinguishes this text-type from handbooks, is the fact that gardening and botanical knowledge is popularised by the author in a chatty, informal way, which combines scientific and professional competence and expertise with personal comments and references to the author's private sphere. These texts seem, then, to reconcile the two areas indicated by Hoyles.

The aim of this chapter is to analyse the main features of this genre, and to evaluate its flexibility in adapting to historical and cultural changes in society. The analysis will be carried out on two case-studies, that is, the gardening notebooks written by two women writers who played an eminent role in British garden literature: *Pot-Pourri from a Surrey Garden* published by Mrs. C. W. Earle in 1897, and *Country Notes* published by Vita Sackville West in 1939. The origin of the genre and its development as well as the influences of other genres on it will also be taken into consideration. Finally, the way private and non-private matters are combined by the authors in what is ostensibly a merely practical collection of gardening notes, will be considered.

2. Gardening Notebooks in the Late Nineteenth-Century: Mrs. C. W. Earle's *Pot-Pourri* Series

The National Library of Ireland holds a small, handwritten volume by Caroline Hamilton, entitled *Garden Notebook*. Caroline Hamilton, née Tighe, of Rosanna in County Wicklow, married Charles Hamilton in 1801. The married couple had an interest in the garden of their eighteenth-century mansion, and planted hedges of yew and beech, as well as roses. Caroline kept a diary, which covered a later period, from 1827 to 1846, in which she listed the works to be done in the garden as well as the plants she planted; she reminded herself of the seeds that had to be bought for her herb garden; she noted remedies against plant pests and diseases, and wrote her observations on garden design and the progress of her own garden (Macleod 1982: 24-28). The notebook was for her a private form of writing, strictly in line with the generic form of a journal. The aims of her garden notebook perfectly match the definition of the notebook, which is in fact "a book reserved for or containing notes or memoranda", as defined by the *Oxford English Dictionary* (1973: 1416) Although, it is still possible to deduce Caroline Hamilton's preferences in terms of flowers (she likes carnations very much), and her spiritual beliefs (she ends her notebook with some religious lines addressing God), the writer of the notebook was writing for her own eyes only, and had a mainly practical aim in mind.

It is in the late Victorian age that we find the first notable published examples of garden notebooks. This is the case of Mrs. C. W. Earle (1836-1925) who was the author of a very successful series of books-*Pot-Pourri from a Surrey Garden* (1897), *More Pot-Pourri* (1899) and *A Third Pot-Pourri* (1903)—which, as their titles metaphorically claimed, were a mixture of interesting information about a variety of subjects connected with house holding and cultivation, because they ranged over vegetable cultivation, herbs and herbals, comments on contemporary gardening books, flower arrangements and literary quotations. Mrs. Earle, who was sixty-one when she published her first book, was an educated woman who had attended the South Kensington School of Art, and belonged to the upper-middle class, which she mainly addressed in her writing. She lived in London in the winter months, but would spend the rest of the year in her house, which was surrounded by a two-acre garden, in Cobham, Surrey.

In contrast with Caroline Hamilton, Mrs. C. W. Earle writes with a readership in mind. The author's note preceding the beginning of the first *Pot-Pourri*, defines the writing as "notes" (2004: 5). This may sound like a disclaimer, but it is really a semantic understatement, "a rhetorical device used by the author to soften possible conflict with the readership" (Chiavetta 2008: 140), and an expression of the identity and subject position she has chosen to meet her readers. Through this text, in fact, Mrs. Earle was getting her active discourse access to the domain of gardening to which male writers had mainly had so far a privileged access. At the same time she was dynamically constructing her identity in discourse, the term "identity" here meant as an active process of discursive work related to other readers and expressing an issue of agency (Benwell/Stokoe 2006: 18). In a similar way, Jane Webb Loudon (1807-1858), who had played an important role in popularising botany for her Victorian middle-class readership of women and children, always used mitigation strategies to win the approval of her readers. In her books, such as *Gardening for Ladies* (1840), *The First Book of Botany* (1841), *The Ladies' Companion to the Flower Garden* (1841), *Botany for Ladies or a popular introduction to the Natural System of Plants* (1842), which were either reference books or handbooks, she never claimed the role of expert, and presented herself as a mediator of knowledge, between the world of the (male) experts, mainly her husband John Loudon who she acknowledged as her teacher, and the world of the nonprofessionals (Chiavetta 2012).

Similarly, in using the word "notes", Mrs. Earle underlines the fact that she is not writing a handbook, and implies that her text is "only" a collection of short, maybe unrelated, pieces. At the same time, she also refers to her choices of genre since, like Caroline Hamilton whose book she had obviously never read, she combines the less formal and scholarly genre of the journal, with the more structured and formal gardening manual. Mrs. Earle's pot-pourri is in fact built on diary entries which not only introduce the tasks to be done in the garden in the various months, but also provide observations and advice on practical matters connected with both gardening and cooking. Mrs. Earle, in fact, enriches her journal with many recipes based on the produce grown in her garden and available in each month. In this way, she hybridised three genres- the journal, the cookery book and the gardening manual, creating a model of her own, which emphasised the domestic and personal quality of her writing. Her books stress, then, how genres are not frozen, but "open to the creativity and indeed transgression of individual agents" (Fairclough 2003: 69), and exemplify what is observed by Vijay K. Bhatia when he analyses genre mixing and bending, which lead to hybridising discourse genres (2004: 87).

Since she mentions the fact that she has kept "a sort of gardening journal for many years, making notes three or four times in the month", in which she would also record any plants bought or brought home from friends "with the date" (Earle 2004: 150), we may consider her choice of genre as an instinctive transfer from a personal, unshared notebook to a shared one. The diary as a text-type retains a quality of privacy, at the same time producing an intimate and involving "I-You" relationship with the readers, who are considered as trustworthy acquaintances of the writer. The pattern Mrs. Earle adopted was so successful- there were eleven reprints of her book in the first year- that she decided to repeat it without great variations in her following *Pot-Pourris*.

As Mrs. Earle often stressed in her books, she was not a professional gardener, but a self-taught amateur. In her first publication, she confesses to her readers that her knowledge about horticultural and gardening matters derives from her extensive reading of gardening handbooks and specialized gardening magazines written by both British and foreign expert gardening writers of the 18th and 19th century. She quotes William Curtis, Joseph Paxton, John and Jane Loudon, and her favourite writer – William Robinson (Earle 2004: 63-84;222-262). Among all these experts, however, she also quotes a writer, whose book, she affirms, had a great impact on her at the very beginning of her learning. The book was a book of flowers, and it was *A Year in a Lancashire Garden*, published in 1879 by an amateur gardener, Henry Arthur Bright. "[…] the book charmed me, and I thought it simple, unaffected, and original", she writes, adding that it was "a fascinating chat about a garden to read in a town and dream over as I did" (Earle 2004: 63). The effect this non-academic book had on her is relevant, because, she affirms, it was while reading it that her wish to have a garden was inspired. This comment on Bright's book

> gives us an idea of the reading and writing relationships of the discourse community of amateur garden lovers which was developing in the Victorian age. This community was united by the awareness of their lack of knowledge in the best of cases, or their unscholarly competence, but also by the pleasure of owning a garden and working in or for it, by the need to find fulfillment there" (Chiavetta 2013: 58).

Bright's text was a "collection of Notes" too (Bright 1981: v), in this case a collection of the articles Bright had written month by month in 1874 for the *Gardener's Chronicle*. In 1875, he had fifty copies of the book printed privately and distributed to friends who shared his passion for gardening. The great success of his work, convinced him that he should have the book published again on a larger scale four years later. The non-scholarly character of his book is highlighted by Bright in his Preface where he affirms that his notes "pretend to little technical knowledge" and are "of but little horticultural value" (Ibid.). He is, however, aware of the pleasure his articles gave to his readers, and of the fact that his readers belong to what today we would define as a discourse community of amateur gardeners. In the Preface he also reports his reluctance to reprint his notes, because he perceives there is "something private and personal about the whole affair" (Bright 1891: vi). At the same time, he acknowledges the fact that "something has been learnt, and something remembered" (Ibid.), by the readers who read his articles in the *Gardener's Chronicle.* The deeply personal nature of his writing is stressed again when he writes "I have described nothing but as it really appeared to me, and I have only wished that others should receive the same impressions as myself" (Bright 1891: vii). At the end of his book, Bright concludes by modestly stating that his aim was merely "to show how much interest and pleasure may be gathered out of a garden of moderate pretensions, and with no great appliances in the way of glass, nor any advantages in the way of climate" (Bright 1891: 87), and he bids a hearty farewell to the readers who have "followed the fortunes", and "shared" with him the hopes of a year in a Lancashire garden (Bright 1891: 88).

Certain key-words in Bright's preface can easily be applied to Mrs. Earle's own *Pot Pourri*, and to most of the notes or notebooks written by gardening writers up to the present day. Sharing one's experiences and pleasure in a garden is the driving force behind these texts. Writers describe and discuss their gardening problems and their solutions, as well as their aspirations, dreams, and the likes and dislikes connected with the piece of land they own and cherish – whatever its size. Readers are therefore introduced into somebody else's garden, sometimes located in a different county and therefore with different soil or climate problems. These problems, however, may be similar, and in this case, the solutions offered by the gardening writer might be adopted by the unknown reader.

To be introduced into somebody else's garden also means being introduced into a private space which creates a bond of intimacy between the writer and the readers. There may be various degrees of intimacy. Bright, for example, is very explicit in expressing what he likes in terms of garden design and flowers to be grown, but nothing much is told about his personal life outside his garden. Mrs. Earle is very determined in expressing her opinions about gardening: "Many gardeners do not agree with me," she declares, "but I am very fond of specimen plants grown in holes cut in grass, if they are planted with care" (Earle 2004: 132). She may be pedagogical, but never patronising, and is always warm-hearted in feeling. "Weeding! What it means to us all! The worry of seeing the weeds, the labour of taking them up, the way they flourish at busy times, and the dangers that come from zeal without knowledge!" (Earle 2004: 105), she exclaims, involving the readers through the use of an inclusive "we". She is also inclined to share her resolute ideas about other topics such as women's education: "Yesterday I paid a visit to the Horticultural College at Swanley," she narrates, "with its branch for women students. It immediately struck me as quite possible that a new employment may be developed for women of small means out of the modern increased taste for gardening" (Earle 2004: 55). Later, she underlines all the positive aspects implied in employing a woman head-gardener "with a man under her, to do the rougher and heavier work" (Ibid.), and also how the "maintaining of a garden and the tending of a greenhouse is a work particularly suited to women of a certain age" (Ibid.). She also tells her readers about her life in general. In her second *Pot-pourri*, for example, she writes about her visit to Florence (Earle 1899: 329-365), and in the third *Pot-pourri* she describes at length her trip to Italy with her son, giving details about her journey and her activities as well as her impressions on Florence, Naples, Pompeii, and Amalfi (Earle 1903: 160-202). The private sphere plays an important part in Mrs. Earle's notes, but this may either be because of the nature of her personality or a deliberate attempt to appeal to her mostly female readership.

Another key concept in Bright's text which is also valid for Mrs. Earle's *Pot-pourri* is the lack of scholarship both authors underline and their insistence on their amateur status. By insisting on their mainly practical competence and above all on the pleasure derived from their gardening activities, both authors delimit the readership they are addressing as well as the boundaries of their work. Their writing springs from their passion for their own gardens and their wish to share. After all, as Timothy Clark notices "the word amateur can be strictly defined from its Latin base, amo – I love, and its English usage someone who works for love" (1980: 76). What Bright and Earle's notes have in common is, then, the informality of their teaching, implicit in the choice of the journal genre. Although their texts are not handbooks, many pieces of information are in fact handed over to readers and many instructions are given. This is done in a direct way, through precise instructions without too many technicalities (e.g. "When they are planted in grass they must, of course, be left alone to take their chance") (Earle 2004: 88), or through advice (e.g. "It blooms better if divided every two or three years") (Earle 2004: 87), or indirectly through the narratives of the author who thus becomes a model to be followed, as in "we take up, from wherever they happen to have been left last autumn, herbaceous Phloxes, early outdoor Chrysanthemums, and

Michaelmas Daisies. These are broken up into small pieces, […] and planted in rows in a half-shady corner of the kitchen garden" (Earle 2004: 89-90).

The yearly rhythm is another feature the two texts have in common. As in most garden notes and notebooks, they follow a seasonal pattern through a monthly order. Notebooks generally start in January and end in December, as in the case of Mrs. Earle, but variations are also possible: Bright, for example, starts his book in December and finishes his year in November.

This monthly ordering was also found in earlier gardening and horticultural literature. In the eighteenth century, an age which saw the development of scientific gardening books, calendars became one of the main genres of gardening writing. As Hoyles notices, "the calendar form is obviously suited to gardening writing because of the importance of timing tasks in the garden throughout the year" (1994: 70). Already in 1664, John Evelyn had published his *Kalendarium Hortense: or the Gardiner's Almanack, Directing what he is to do monthly throughout the year, and what fruits and flowers are in prime*, a handbook in which he "allocated tasks to each month, with a rather lyrical, swinging approach to the job" (Uglow 2005: 108). In 1732, Philip Miller published *The Gardener's Kalendar*, which had sixteen editions in fifty years. The aim of the book was practical, since it set out what had to be carried out in the garden in each month. In this text, however, there was no space for personal private observations. Another example was *Beeton's New Dictionary of Every-Day Gardening* published in 1896, (a reissue of 1895's *Beeton's New All About Gardening),* which was "rendered the more valuable and helpful by the monthly calendar of garden work throughout the year" which exhibited a "complete Compendium of all that should be done in each and every stage of Gardening through the year's successive months" (Beeton 1909). Once again, as expected in a text added to an encyclopaedia, there was no space given to any comment other than practical and direct instructions.

What made Mrs. Earle's notebook different from these calendars is the fact that, still respecting a cyclical pattern which took monthly tasks into consideration, she displayed wider interests. Her "unique amalgam of garden lore, cookery hints, reminiscences, suggestions for further reading, and much else" (MacLeod 1982: 33) showed how flexible the genre could be. She also added a personal touch to the garden notebook, showing her readers how a gardening text could be considered an instrument to express oneself and communicate something more than mere instructions or advice.

Gardening Notebooks in the 1930s: Vita Sackville-West's *Country Notes*

As a gardening writer, Vita Sackville-West is mostly famous for her series of books, *In Your Garden* (1951), *In Your Garden Again* (1953), *More for Your Garden* (1955), *Even More for Your Garden* (1958). These were collections of the articles she had published in the *Observer*, *The New Statesman* and other newspapers or magazines. A few decades before these successful writings, she had written two other gardening books – *Some Flowers* (1937), a collection of short essays in which she describes twenty-five of her favourite flowers, and *Country Notes* (1939).

Country Notes is a collection of the articles she had published at fortnightly intervals during 1938 and 1939 in the *New Statesman and Nation*, and of the articles she had written for *Harper's Bazaar*, *Vogue*, *Time and Tide*, *Travel*, and *The Nation and Athenæum*. The book is enriched by many black and white photographs of rural and country life. The definition of "notes" to describe these texts appears in the acknowledgments, and possibly depends on their shortness since they are generally one or two pages long. Since the text is formed by magazine articles, the book is an example of scientific journalism, and the author may be considered a populariser addressing a discourse community of both experts and non-experts who share an interest in or passion for the topic of nature. However, although some articles give pieces of advice on gardening matters, the main objective of the writer is to converse with her readers in an informal way on various topics connected with country life. The scope of Sackville-West's notes is as wide as Mrs. Earle's since it is not limited to gardening issues, but considers country life at large. There are no recipes or household advice, but many literary references are present, which does not surprise since Vita Sackville-West was a renowned poet and novelist.

In *Country Notes*, the traditional calendar pattern is followed, with the text organized in monthly order, even though only a few of the articles have the name of the month included in their title. After the introductory note ("Country Life"), for example, there is a note entitled "January'" followed by "Winter Colour" and "February Frost", but the next month mentioned in the title is "May", while the notes in between which focus on springtime have different titles, though there are internal references to March and April.

Although the genre here is the feature article, and not the diary, as was the case with Mrs. C. W. Earle's writing, sometimes Sackville-West refers to her private garden-diary and quotes her personal journal entries to reconstruct what happened to her flowers during the year. "I find that on January 26th the blue primroses were in full flower, thus preceding their ordinary flowering period by about two months", she writes, adding that she finds a note on March 9th saying "all primroses flowering in earnest"; another note at the end of March is also quoted, which reminds her of the profusion of tulips, hyacinths, anemones and irises in her garden. As in the case of Mrs. Earle, there is then a link between her private diary and the published work.

The first note, "Country Life" introduces the author's perspective on life. The note focuses on the contrast between the author's happy country life and the life of her friends living in London. Sackville-West fills this first text with thoughts and queries provoked by the visits paid to her by her metropolitan friends. On the one hand, there is her "rustic peace", and on the other "the reverberation of a wider world" they bring with them (Sackville-West 1939: 11). "They ask me if I have seen this or that play, these or those pictures", she comments, "and always I find myself obliged to reply that I have not. This makes me appear, and feel, a boor" (Ibid.) Readers are allowed access to the intimate reactions of the writer, who shares with them both her feeling "slightly disturbed", when the friends go back to their city life, and her "insulting sense of calm superiority" (Ibid.). The rhetorical questions that follow still deal with the contrast between an active, agitated life, concerned with all the troubles of the world, and an attitude of mind which pays attention to nature's little events, which are valued as a "deep source of delight" (Ibid.). On the brink of the Second World War, Sackville –West underlines not only the pleasure, but also the healthy power of country life, which shows us every day "the eternally renewed evidence of the determination to live" (Ibid.). Although she is describing the constant renewal of nature's cycle, a moral statement is affirmed by the author which could certainly be reassuring for her readership at that time, while also emphasizing the importance of rural Britain values: "The powers of resistance against adversity are greater than we thought; the germ of life lies hidden in the midst of apparent death", she declares, adding that she welcomes "the youth of the new season, whether it comes with the first lambs to be born in a snow-storm or with the new buds of the hornbeam pushing the old brown leaves off the hedges" (Sackville-West 1939: 11-12). The final sentence of the note addresses her readers directly, thus involving them in her train of thoughts and, at the same time winning their agreement: "if you have a taste for such things, no amount of repetition can stale them; they stand for permanence in a changing world" (Sackville-West 1939: 12). Vita's philosophy of life is thus made clear through these initial pages which may be considered a manifesto on which all the other notes will be built. This philosophy will be consistently presented throughout the book, while the author deals with the various features of country life.

In *Country Notes*, Sackville-West describes and narrates what happens to her while living in the country; she introduces her habits, her decisions; she mentions her friends and the routine of her life. She always expresses her opinions and beliefs in a strong and determined way, even when she asks herself (and therefore her reader) many questions which remain unsolved. The reader, thus involved, easily accepts the fact that Sackville-West is the real protagonist of the articles, and that entering her world is a precious and instructive experience. She becomes a sort of opinion leader. “I do not like January very much” (Sackville-West 1939: 14), she states, and gives her reasons for this dislike, and in another article on gardening, she affirms she finds gardeners “disconcerting people”: “either they know infinitely more about the subject than I do, or else they know infinitely less. Seldom do I encounter one with whom I can discuss our common topic on equal terms” (Sackville-West 1939: 109).

The personal tone of Sackville-West’s notes allows a wide space for memories, often from her childhood or youth. The one-page note, entitled “Shooting”, for example, tells of her going shooting with her father. This dense narration recreates not only the image of a gentleman who did not enjoy “big covert drives”, but preferred “rough shooting, where he felt that he and the birds stood a more even chance”, but also reveals a father-daughter relationship, in which the two shared the same attitude to life-“Freedom was ours and the sporting chance”, she writes (Sackville-West 1939: 84). At the same time, Sackville-West portrays an aristocratic, fashionable setting to which she used to belong: “On those richly golden autumnal afternoons, with his loader and dog, we must have looked like a photograph in the social illustrated papers”, she comments, adding that she does not think “he enjoyed it any more than I did” (Ibid.).

She also sketches her biography as a gardener. “I myself took to gardening quite late in life”, she writes, “I must have been at least twenty-two” (Sackville-West 1939: 96). She then recollects memories of her negative attitude to gardening as a child, when she was given “a strip which was known as ‘Vita’s garden’, because of the tradition that every child must automatically love and cherish a garden of its own” (Ibid.). This was a Victorian tradition she disagrees with, since, she believes, children are too unmethodical and impatient to enjoy gardening. There are references to Knole, the Elizabethan family manor in which Vita was born (“As a child I had the good fortune to live in an exceptionally beautiful home, with acres of garden to match all enclosed within a light stone wall") (Sackville-West 1939: 97), and a humorous description of the head-gardener, "an immensely dignified man, with a hooked nose, keen eyes, and a great black beard, giving him the appearance of a major prophet", who was "the terror of [her] life" (Sackville-West 1939: 97). She then expresses her preference for the second-in-command, "a gentle timid little man who had a real feeling for flowers" (Sackville-West 1939: 99), a born gardener who never scolded little Vita for her misdeeds in the garden, but "only pleaded", thus teaching the child how to respect and love plants. "He explained to me so wistfully why he minded my cutting the delphiniums he had just spent hours in staking that I felt ashamed of my thoughtlessness" (Ibid.), she confesses. This article, entitled "Gardens and Gardeners", is one of the longest in the collection. Through its descriptions of gardening life and tendencies, Sackville-West recreates an Edwardian aristocratic atmosphere, portraying her family background and her own upbringing. In the same article, she also analyses, with the experience she has by now acquired, the old-fashioned gardening style of Knole's head-gardener, and compares it with the innovative ideas of William Robinson, who "did more to alter the fashion of English gardening than any man of his time" (Sackville-West 1939: 106), and whose ideas she shares. She tells how she met the old gardening theoretician and writer, and her description offers a delicate portrait of the aged rebel at the end of his life: a "courteous old man[…] in his invalid-chair, sitting out in the sun among his flowers, unable to raise his head, scarcely able to move his hands, but still sufficiently alert to take pride in the garden he had created and to talk with knowledge and enthusiasm about his schemes and his innovations" (Sackville-West 1939: 105). Thus, by allowing her readers to share her memories, Sackville-West also reconstructs a national gardening heritage which belongs to each of her readers. In other articles, she shows she is a keen historian: some pages are dedicated to the history of the hop during the reigns of Henry VI, Henry VII and Henry VIII;other pages focus on coppice planting, which dates back to 1545. Pride in communal ties is thus strengthened between her and her readership, who are already united by their passion for gardening. In many pages, she builds a positive image of this community of gardeners, who are endowed with many qualities. They are "always optimistic, always enterprising, and never satisfied", she states. "They always look forward to doing better than they have done before" (Sackville-West 1939: 174); they are "the few people left in this distressful world to carry on the tradition of elegance and charm" (Sackville-West 1939: 173).

Meta-discourse is also present. Being a reflective writer, Vita Sackville-West comments on the difficulty of writing about nature in an unsentimental and sincere manner. "If I set down that I have to-day seen apple-blossom strewn by wind on grass," she affirms, "I am stating a fact, and if I should happen to re-read my own words in future years (which is unlikely) they will probably recall that vision, as fresh and bright in memory as on that morning in the month of May" (Sackville-West 1939: 75). On the other hand, she continues,

> If[…] I start to expand my statement, in the hope of evoking a similar vision in the mind's eye of another, I shall immediately find myself drawn into semi-falsities, into truth wrapped round with untruth; I shall begin to search for what the apple-blossom was like; I shall find confetti or snowflakes as a convenient comparison; I shall hit on the words shell-pink to express the delicacy, the papery delicacy of the scattered petals; I shall begin 'to write'; but really, if I can be sufficiently severe with myself, I shall put my pen through all those blobs of ink, those wordy words, and cut myself back to the short phrase about apple-blossom strewn by wind on grass (Sackville-West 1939: 75-76).

"Writing is indeed a strange and difficult profession", she finally confesses (Sackville-West 1939: 76). Thus, by talking about the process of writing, which plays such a great part in her personal life, Vita allows her readers to be in touch with a very personal aspect of her personality.

Country Notes is an interesting combination of poetic descriptions, resolute statements, historical information and, of course, practical suggestions about gardening issues. Very few instructions are given directly to the readers; in most cases, there are suggestions or accounts of her own behaviour in the garden. There is a long description of the hop-picking season (Sackville-West 1939: 144), and of oast, and the process from "bine to bottle" (Sackville-West 1939: 148). She also talks about Sissinghurst, her garden in Kent, and about the pleasure of having other gardeners visiting it. Like Mrs. Earle, she also includes some notes about her trips abroad – to Italy and France, once again showing how a genre can be adapted and expanded.

In many pages, Sackville-West portrays herself determined to keep her distance from the world outside her life in the country. She does not deny the importance of the events that soon will overwhelm her world too, but she patently shows that she prefers to detach herself. In "One Afternoon", for example, she starts mentioning that at 3 p. m. she was listening to "the first report of Herr Hitler's speech to the Reichstag", but that by 3. 30 p. m. she had "gotten [herself] with relief out into the very different atmosphere of the open fields, the quietly busy fields, busy with their April life. Corn was springing, larks were singing, lambs and piglets being born." (Sackville-West 1939: 67). Her loving description evokes an idyllic atmosphere, threatened by Hitler's speech, in which she finds refuge. The article then introduces another character – Henry, the man who looks to the sheep and pigs. He becomes a figure to contrast with Hitler. Henry deals with the birth and death of the animals he looks after in a detached and practical way, but he is also patient and gentle. He represents the rural virtues of England, a centuries-old expertise, acquired through experience handed down from one generation to the other, so much so that the writer starts thinking of the herdsmen of Elizabethan times.

Through the character of Henry, Sackville-West portrays her own world for the readers, and her rhetorical questions once again introduce thoughts and queries, which many other people must have shared with her in those difficult days:

> Now why, I often ask myself, should these ordinary things give me so deep and lasting a satisfaction? Why should Henry's simple expressions move me more profoundly than any Dictator's rhetoric? Which, after all, is likely to affect my life and the lives of my countrymen in far greater degree. (Sackville-West 1939: 70).

In a poetic, metaphorical language, the author admits she is perfectly aware of the limits of her attitude. "I sometimes think" she writes, "that the love of nature and the natural seasonal life may attain the proportion of a vice; may obsess one to the extent of desiring nothing else, nothing beyond: a drowning, a lethargy, an escape, an indolence and an evasion" (Ibid.).

Despite this tentative self-accusation, patriotic feelings are nevertheless present in these pages. It is in *Country Notes* that we find the famous definition of Great Britain as "a nation of gardeners" (Sackville-West 1939: 85), and the description which follows stresses a vital difference between Britain and Germany, seen through their life styles and gardening attitudes: "A nation that so profoundly and extensively loves flowers and dogs", the author comments, "must surely have something very unbellicose in its make-up" (Ibid.). She continues by recalling a flower-show which she had attended in Berlin and through her words the exhibits become symbols of dangerous war weapons ("bayonets", "pikes") and of an inexorable violent death. We see how aware she is of the horrific signals of war and death which will soon overshadow the lives of British people, but we also see in this description how she values and cherishes the graciousness and amiability of the people of her own country:

> the principal exhibit was of the most spiky and truculent forms of cactus, grey-green bayonets, murderous pikes – a horrid symbol, I thought, at the time, although that was more than ten years ago and the Führer's name

> only an insignificant whisper. The second principal exhibit was an avenue of tombstones, appropriately decorated with wreaths or bouquets of tin flowers. In a flash I saw English cottage gardens; I saw the fortnightly shows at Vincent Square; the dove-coloured spinsters coming up from the country with their little note-books, peering closely down into the mealy auriculas, taking notes, placing their orders rather cautiously[…] then lingering over the flaming azaleas, the rose-red prunus, all so enticing; […] they are dealing with real possible flowers, not with the sinister cacti and mortuary wreaths of the Berlin equivalent of our Chelsea show (Sackville-West 1939: 85-86).

The gardening notes become, then, a way to express much more than instructions, suggestions or personal observations. They assume an ideological meaning as they turn to a way to declare the belonging to one's homeland and the willing acceptance of its principles.

CHAPTER THREE

STANCE IN POPULARISATION: INTERACTION OR MANIPULATION? DARWIN'S THEORY OF EVOLUTION AND RELATED MATTERS AS PRESENTED IN GEORGE WILLIAM HUNTER'S *A CIVIC BIOLOGY PRESENTED IN PROBLEMS*(1914)

THOMAS CHRISTIANSEN

1. Introduction

In this chapter, we investigate the language of William Hunter's discourse in the famous high school textbook *A Civic Biology Presented in Problems*(1914) concentrating on speaker stance (Biber *et al.* 1999; Hunston and Thompson 2000) including the interpersonal function, and in particular the way in which *ethos* (Baumlin 1994; Cherry 1998; Johnstone 2009) features in the presentation of Darwin's theories. Comparison will be made with those works of Darwin's dealing with evolution. The aim will be to show how and where Hunter injects his own stance into Darwin's theories thereby illustrating not only the interactive aspect of popularisation but also its retrogressive aspect, namely the way that ideas may be mistranslated (i.e.distorted and misrepresented) and thus scientific "progress" checked.

2. Popularisation and Pedagogy

The process of popularisation, whereby specialized knowledge (e.g. scientific, technical) is diffused to a wider general audience lies at the heart of education although it is useful to draw a distinction between popularising and pedagogic texts, because, although the latter may be included in the former, pedagogic tests have the additional aim of initiating the reader into the "secondary culture" (Widdowson 1979):

> [p]opularizations target instead a wide reading public and deal with specialized topics in a language close to general discourse and to the layman's everyday experience. The purpose here is chiefly informative and seeks to extend the reader's knowledge rather than develop a secondary conceptual system (Gotti 2013: 10).

Texts like Hunter's aimed at public high school education (to use the appropriate term in its US context) display features both of popularisation and pedagogic texts in that it can be seen that their object is primarily to inform in a way that relates to the students' immediate needs and daily lives and not necessarily to turn them into biologists as made clear in Hunter's introduction:

> In a course in biology the difficulty comes not so much in knowing what to teach as in knowing what *not* to teach. The author believes that he has made a selection of the topics most vital in a well-rounded course in elementary biology directed toward civic betterment. The physiological functions of plants and animals, the hygiene of the individual within the community, conservation and the betterment of existing plant and animal products, the big underlying biological concepts on which society is built, have all been used to the end that the pupil will become a better, stronger and more unselfish citizen (Hunter 1914: 10-11).

In Christiansen (2013), a model of popularisation is proposed that rejects the traditional view that the process of popularisation is essentially one of a one-way simplification that occurs after a theory has been formulated by the expert, the *savant*, and delivered to the non-expert, seen as the *ignorant*[1]much as if the latter were a consumer of pre-packaged ideas. Instead it is posited that, especially in the case of Darwin (a private scholar, not a tenured academic, for whom correspondence with experts and non-experts alike was a vital part of his research-see Christiansen 2010), popularisation constitutes an exchange of ideas whereby theories, ideas and concepts are aired and revised or elaborated according to the

[1] See Raichvarg and Jacques (1991).

reception that they receive. From this perspective, the expert, though still in a privileged and dominant position, is not exclusively active nor is the receiver entirely passive.

Such a model can also be said to be true of a school textbook not so much in the fact that students have the power or opportunity to engage in debate with the author (which of course they typically do not) but rather in the process by which such books are commissioned and adopted. In the USA in particular, school boards are elected bodies and the content of lessons and teaching material is open to public scrutiny and may become a political (or religious) issue. It may not then simply be a case of *savants* and *ignorants* but rather of opposing ideologies, in which the expert's dominant scientific position is irrelevant, at least in the short term.

This is certainly the case with Darwin, who is as much derided by some as he is respected by evolutionary biologists, especially within the context of the various states of the USA, education being rigorously a local affair. Hunter's *A Civic Biology Presented in Problems* (1914) -henceforth *Civic Biology* or CB achieved a fame / infamy for being at the centre of the historic Scopes Trial of 1925[2]. The fact that CB became a standard bearer for modern science in the struggle with religion and tradition is somewhat ironical because, in that which was billed at the time as the epic showdown between the Bible and Darwin, on many issues, Hunter is hardly a fair representation of Darwin's position at all. Concentrating on constitutional and religious issues, this was a point entirely missed by observers during the trial. Indeed, not wishing to side with creationists on a distinctly scientific matter (and there is no reason why Darwinism should be seen incompatible with religious teachings) [3], it is noteworthy that CB has since achieved a kind of notoriety because, specifically in the parts purporting to present Darwin's theories, it

[2] Formally *The State of Tennessee v. John Thomas Scopes* (1925) where a high school teacher was put on trial for teaching those sections of Hunter pertaining to Darwin's theory of evolution in defiance of Tennessee's Butler Act, forbidding the teaching of anything contradicting biblical accounts of the origins of humankind in any state-funded school. Scopes was acting on political motives and had been encouraged by the *American Civil Liberties Union* who were determined to put the law to the test (he even admitted, much later, that he may even have not taught from the sections in question). Initially found guilty, the case was thrown out on a technicality. The effect of the trial was that textbooks usually avoided evolution until after the 1968 Epperson v. Arkansas decision. Evolution continues to face opposition and remains a controversial issue in the USA especially with proponents of so-called "creationism" and its pseudo-scientific cousin "intelligent design".

[3] See Moore (2002).

espouses an extreme form of evolutionary biology that is abhorrent to most modern readers.

This will also present in itself an interesting case study in the process of popularisation and its wider scientific and cultural implications, in particular the ways in which misinterpretation and misappropriation may affect the process of popularisation thus not only underlying the interactive nature of this process as described by Christiansen (2013) but also how the process itself may be manipulated for ideological reasons. To do this we must examine the text from the viewpoint of stance, which we will discuss in Section 4. Before doing so, in the next short section we will outline the corpus of texts that we analyse.

3. Corpus

To conduct this study we compared *Civic Biology* with those works by Charles Darwin (Figure 3. 1.), who features prominently in Hunter, where evolution and related matters were explicitly discussed. These were identified by searching for items from a list of terms relating to evolution / creation identified by Christiansen (2009) found in a larger corpus of Darwin's published works and published correspondence (Table 3.1.):

distinct creation	separately created	
distinctly created	evolution	preserved naturally
created distinctly	evolve (s)	naturally preserved
independent creation	evolved	sexual preservation
independently created	evolving	sexually preserved
created independently	natural selection	preserved sexually
individual creation	naturally selected	descent with
individually created	selected naturally	modification
created individually	sexual selection	hereditary modification
multiple creation	sexually selected	inherited modification
separate creation	selected sexually	survival of the fittest
created separately	natural preservation	law (s) of inheritance

Table 3.1. Terms in Darwin's writings relating to evolution / creation (Christiansen 2009).

By using the terms listed in Table 3.1. as a filter, we were able to identify the published works of Darwin in which evolution or matters related to it are specifically discussed. Where different editions exist (e.g. *On the Origins of the Species*), we took the last edition, to avoid duplication of results. This comparison corpus consisted of 12 works (total

1, 822, 083 words). In Table 3.2., we summarise the composition of the corpus.

<table>
<tr><td colspan="2">George William Hunter's Civic Biology (124, 884 words)</td></tr>
<tr><td colspan="2">Civic Biology foreword (2211 words)
Civic Biology minus evolution (110, 896 words)
Civic Biology evolution (11, 777 words):
• XIV. Division of Labor, The Various Forms of Plants and Animals" (pp. 173-196);
• XVII. Heredity, Variation, Plant and Animal Breeding" (pp. 249-265);
• XXV. Some Great Names in Biology (pp. 398-406).</td></tr>
<tr><td colspan="2">Darwin's academic works in which matters related to evolution are discussed(1, 822, 083 words):</td></tr>
<tr><td>The Movements and Habits of Climbing Plants (1865-1882)
The Descent of Man (1871)
Expression of Emotion in Man and Animals (1872)
On the Origin of the Species (6^{th}ed 1872)
Insectivorous Plants (1875)
Animals and Plants under Domestication Vol I (2^{nd}ed 1875)
Animals and Plants under Domestication Vol II (2^{nd}ed 1875)</td><td>The Power of Movement in Plants (1875)
The Effects of Cross and Self-Fertilization in the Vegetable Kingdom (1876)
The Different Forms of Flowers and Plants of the Same Species (1877)
On the Fertilization of Orchids (2^{nd}ed 1877)
The Formation of Vegetable Mould through the Action of Worms with Observations on their Habits (1881)</td></tr>
</table>

Table 3.2. General data on corpora.

Within *Civic Biology*, we distinguished between those sections dealing with evolution (CB evolution) and related matters and the other sections of the work CB minus evolution. We also categorised the foreword as a separate section because the author addresses the reader directly about the work itself, not the subject matter, biology.

4. Stance

Stance is a very general concept relating to the position as regards the discourse taken by a given participant. According to Kockelman (2004: 127), stance may be understood as the semiotic means by which an addresser indicates her[4] orientation to states of affairs, usually framed in

[4] Following Sperber and Wilson (1986), we use the feminine for addressors and masculine for addressees.

terms of evaluation (e.g. moral obligation and epistemic possibility) or intentionality (e.g., desire and memory, fear and doubt). In reviewing the literature, Kockelman identifies four fundamental types of stance:

- *Social (Ethical) stance* = ideological standpoint (e.g. "political correctness").
- *Authorial stance* = positioning of addressor within and towards text.
- *(Inter) personal stance* = perspective of participant.
- *Intentional stance* = interpretation of social behaviour in terms of putative mental states (e.g. belief, desire, fear).

Of these, the most important in the context of CB are the first two: the social / ethical stance and authorial stance. The last two, though not entirely absent, are features more typical of interactive discourse than of a textbook. In the sections below we shall discuss firstly how Hunter's approach to evolution can be seen to differ from Darwin's and reflect what are, we can suppose, his own views on the subject, even though he presents them as Darwin's. Secondly, we will analyse how he establishes his persona as author, which allows him to set himself up as an authority in a position to evaluate and pass judgment on the various theories of evolution.

4.1. Ideology and Social / Ethical Stance in *Civic Biology*

To examine how Hunter's social stance is manifested in CB, we must first briefly discuss the scientific impact of Darwin's theory of evolution—the reason why it was to be popularised in a textbook—and also its political and cultural ramifications, which accounts for the different ideological positions taken in regard to it, which culminated in the Scopes Trial.

Darwin's *Origin of the Species*, his work most closely associated with evolution, was an immediate commercial as well as scientific success. All 1, 250 copies of first addition sold out on the first day, a remarkable feat for a scientific work, and five further editions were produced over the next 13 years. By 1869, according to Hull *et al.* (1978), nearly 75 percent of all scientists had accepted Darwinian evolution. In the wider community, Darwin's theories constituted a turning point where old certainties about religion and humankind's place in the world were profoundly shaken with far-reaching effects in all areas of learning, culture and life (see Glick 2010).

Within the scientific community, by the early 1900s, Darwinism had however suffered an "eclipse" (see Huxley 1942; Bowler 1983), when the concept of evolution was widely accepted but his account of natural selection was not considered an adequate explanation for it (see Christiansen 2012). It was not until the so-called *modern evolutionary synthesis* (Dobzhansky 1937) when research from other areas of biology, in particular quantitative and population genetics (Hunt Morgan 1915; Fisher 1930; Wright 1932; Haldane 1932-all developments of Mendel's methods), helped to provide a more robust explanation of how descent with modification functions that evolution was widely accepted again in scientific circles.

In the 1910s and 20s, evolutionary biology was therefore in a kind of limbo. There was broad agreement that Darwin was on to something but it was recognized that he was unable to explain how evolution worked. Within this vacuum, several increasingly less scientific and more unsavoury ideas and theories were spawned: "survival of the fittest"[5]; pangenesis (inheritance of acquired traits); social Darwinism (Schmidt 1877); eugenics (Galton 1883, Darwin's half-cousin). Hunter's approach to evolution is found among these, although before being too critical of him, it must be assumed that his views were shared by many as evidenced by the fact that the prosecution in the Scopes Trial never took his textbook to task for its somewhat extreme take on Darwin's theory of evolution. In Table 3.3. we illustrate the distance between Hunter and Darwin on key points by contrasting often cited sections from Hunter (whose original titles we retain) with passages from Darwin (one from *Origin*, three from *Descent of Man*) dealing with the same issues:

---

[5] The expression "survival of the fittest" was first used by the philosopher Herbert Spencer (1864) and proved to be the central concept of Social Darwinism. Darwin adopted the term in the 5th edition of *Origin of the Species* as a useful shorthand for natural selection despite the reservations of some naturalists, notably Arthur Wallace.

HUNTER	DARWIN
HUMAN EVOLUTION Undoubtedly there once lived upon the earth races of men who were much lower in their mental organization than the present inhabitants. If we follow the early history of man upon the earth, we find that at first he must have been little better than one of the lower animals. He was a nomad, wandering from place to place, feeding upon whatever living things he could kill with his hands. Gradually he must have learned to use weapons, and thus kill his prey, first using rough stone implements for this purpose. As man became more civilized, implements of bronze and of iron were used. About this time the subjugation and domestication of animals began to take place. Man then began to cultivate the fields, and to have a fixed place of abode other than a cave. The beginnings of civilization were long ago, but even to-day the earth is not entirely civilized. (pp. 195-196)	I should premise that I use the term Struggle for Existence in a large and metaphorical sense, including dependence of one being on another, and including (which is more important) not only the life of the individual, but success in leaving progeny. (*On the Origin of the Species* p. 62) The following proposition seems to me in a high degree probable-namely, that any animal whatever, endowed with well-marked social instincts, the parental and filial affections being here included, would inevitably acquire a moral sense or conscience, as soon as its intellectual powers had become as well, or nearly as well developed, as in man. For, firstly, the social instincts lead an animal to take pleasure in the society of its fellows, to feel a certain amount of sympathy with them, and to perform various services for them. (*Descent of Man p.* 71-72)

IMPROVEMENT OF MAN If the stock of domesticated animals can be improved, it is not unfair to ask if the health and vigor of the future generations of men and women on the earth might not be improved by applying to them the laws of selection. This improvement of the future race has a number of factors in which we as individuals may play a part. These are personal hygiene, selection of healthy mates, and the betterment of the environment. (p. 261)	With highly civilized nations continued progress depends in a subordinate degree on Natural Selection; for such nations do not supplant and exterminate one another as do savage tribes. Nevertheless the more intelligent members within the same community will succeed better in the long run than the inferior, and leave a more numerous progeny, and this is a form of Natural Selection. The more efficient causes of progress seem to consist of a good education during youth whilst the brain is impressible, and of a high standard of excellence, inculcated by the ablest and best men, embodied in the laws, customs, and traditions of the nation, and enforced by public opinion. It should, however, be borne in mind, that the enforcement of public opinion depends on our appreciation of the approbation and disapprobation of others; and this appreciation is founded on our sympathy, which it can hardly be doubted was originally developed through Natural Selection as one of the most important elements on the social instincts. (*Descent of Man* p. 143)

EUGENICS When people marry there are certain things that the individual as well as the race should demand. The most important of these is freedom from germ diseases which might be handed down to the offspring. Tuberculosis, syphilis, that dread disease which cripples and kills hundreds of thousands of innocent children, epilepsy, and feeble-mindedness are handicaps which it is not only unfair but criminal to hand down to posterity. The science of being well born is called eugenics. (p. 231)	With savages, the weak in body or mind are soon eliminated; and those that survive commonly exhibit a vigorous state of health. We civilized men, on the other hand, do our utmost to check the process of elimination; we build asylums for the imbecile, the maimed, and the sick; we institute poor-laws; and our medical men exert their utmost skill to save the life of every one to the last moment. There is reason to believe that vaccination has preserved thousands, who from a weak constitution would formerly have succumbed to small-pox. Thus the weak members of civilized societies propagate their kind. No one who has attended to the breeding of domestic animals will doubt that this must be highly injurious to the race of man. (*Descent of Man* p. 133-134)

PARASITISM AND ITS COST TO SOCIETY Hundreds of families such as those described above exist today, spreading disease, immorality, and crime to all parts of this country. The cost to society of such families is very severe. Just as certain animals or plants become parasitic on other plants or animals, these families have become parasitic on society. They not only do harm to others by corrupting, stealing, or spreading disease, but they are actually protected and cared for by the state out of public money. Largely for them the poorhouse and the asylum exist. They take from society, but they give nothing in return. They are true parasites. The Remedy. –If such people were lower animals, we would probably kill them off to prevent them from spreading. Humanity will not allow this, but we do have the remedy of separating the sexes in asylums or other places and in various ways preventing intermarriage and the possibilities of perpetuating such a low and degenerate race. Remedies of this sort have been tried successfully in Europe and are now meeting with some success in this country. (p. 263)	The surgeon may harden himself whilst performing an operation, for he knows that he is acting for the good of his patient; but if we were intentionally to neglect the weak and helpless, it could only be for a contingent benefit, with an overwhelming present evil. We must therefore bear the undoubtedly bad effects of the weak surviving and propagating their kind; (*Descent of Man* p. 134)

Table 3.3. Comparison of notorious quotes by Hunter with selected citations from Darwin.

As can be seen, although Darwin at times seems to espouse eugenics (see his comments on this subject), his support is at best tentative, almost reluctant. It is true that when he associated himself with expressions like "survival of the fittest" and initially showed enthusiasm with the efforts to prove the existence of so-called *gemmules* or *pangenes* (the means by which acquired, not inherited, characteristics could be passed on) by his half-cousin Francis Galton (see Christiansen 2012), who was later to become a leading figure in the field of eugenics, Darwin had given some, certainly unintentional encouragement to the proponents of so-called Social Darwinism. As the selected quotes in Table 3.3 make clear, although Darwin was to suffer a crisis of faith in his later life, and was aware that the laws of evolution, as he saw them, could lead to an essentially survivalist view of society and existence in general, he did not believe that biology should take precedence over morality. He even tried to show how sympathy towards others and protection of the weak was not irreconcilable with evolution. He believed that such sentiments may indeed be the product of natural selection and are what distinguish humans from "lesser" creatures (see comments on human evolution and improvement of man).

By contrast, Hunter embraces the extreme versions of eugenics and Social Darwinism wholeheartedly. The latter in particular, as the name itself makes abundantly clear, appropriates Darwin's theories for its own agenda, lending support to some distinctly unsavoury attitudes towards certain sections of society (e.g. the "weak", the "infirm", the "racially inferior"). Whether by ignorance or guile, Hunter takes this misrepresentation a step further by voicing such opinions uncritically in a textbook for schoolchildren, attributing them not to their real authors such as Spencer (a philosopher / economist), Galton or Schmidt, but portraying them as belonging to the leading and most respected natural historian of the time.

We can illustrate this point by listing in Table 3. 4. the sources cited by Hunter with those cited in Darwin's works dealing with evolution:

	Darwin	Galton	Lamarck	Mendel	Schmidt	Spencer
Civic Biology	15. 12	0. 85	0	5. 37	0	0
Darwin Evolution	0. 25	0. 21	0. 07	0	0. 01	0. 29

Table. 3.4. Notable evolutionary scientists cited in *Civic Biology* compared to corpus of Darwin's works dealing with evolution.

In Table 3.4., results are weighted accorded to size of corpus e.g. the figure for Mendel in CB of 5. 37 is the number of actual occurrences (nine) divided by 124, 884 (i.e. total words in corpus), multiplied by 100. The latter an arbitrary figure to avoid having figures so small that the number of initial zeros make them difficult to display and work with. As can be seen, Darwin cites a wider selection of notable scientists than does Hunter but cites each much less. [6] Hunter cites only three (including

[6]In his works, Darwin is always assiduous in giving all his sources quoting even personal correspondence with enthusiasts, papers read to small meetings as well as the works of his peers in the academic community around the world.

Darwin) but much more for each. Most importantly, Hunter does not cite Schmidt or Spencer at all, although he gives due importance to Galton. He does however cite Mendel showing that he was surprisingly up to date with the most recent developments leading up to the modern evolutionary synthesis (see above). [7] Darwin is, however, by a wide margin, presented as the primary source for Hunter's sections on evolution.

In Table 3.5. below, we look at the terminology related to evolution employed by Hunter, comparing it to that found in Darwin's works related to evolution (see Table 3. 1.):

	Civic biology (124, 884 words)	Darwin Evolution (1, 822, 083 words)
descent with modification	0	0. 15
distinct creation etc.	0	0. 01
evolution etc.	1. 6	0. 4
hereditary modification etc.	0	0. 01
independent creation etc.	0	0. 13
inherited modification	0	0. 01
law (s) of inheritance	0	0. 19
natural preservation etc.	0	0. 02
natural selection etc.	0. 08	4. 09
separate creation etc.	0	0. 05
sexual selection etc.	0	1. 82
"survival of the fittest"	0	0. 15

Table 3.5. Terms related to evolution from Darwin featuring in *Civic Biology*.

Unsurprisingly for a school text book, CB uses far less terminology than the original academic work which it purports to discuss. Darwin uses a whole range of terms to describe the mechanism of evolution, many of

[7] Darwin was unaware of Mendel works even though the latter had published his work on genetics just two years (1866) before Darwin's epic treatment of the subject of inheritance: *Variation of Animals and Plants Under Domestication* (1868). Mendel's work was all but ignored until after 1900, and in particular after 1915, when his laws of *Segregation* and of *Independent Assortment* were integrated with the chromosome theory of inheritance by Thomas Hunt Morgan.

which are synonymous: something untypical of specialized discourse as it does not adhere to the principle of *monoreferentiality*, not to say *precision* (Gotti 2003). Furthermore, the fact that Darwin employs so many different terms to describe essentially the same thing is because at the time of writing he is unsure of the exact nature of what he is discussing and is experimenting with different descriptions. *Evolution* ("evolved", "evolving" etc.) is a term he initially uses less than variations of *modification* and it only features prominently in later editions of Origins (see Christiansen 2009). Indeed, the figure 0. 4 is still much lower than *natural selection* (4. 09) etc. or *sexual selection* (1. 82). Hunter, by only using *evolution*, is largely following what had become convention among biologists in 1914. The other term he uses is *natural selection*, which relates to the cause of evolution.

Surprisingly, despite the social Darwinist views expressed by Hunter (Table 3.1.), the expression "survival of the fittest" does not feature at all in CB. This is interesting as this simple term sums up not just how evolution occurs but also, for the social Darwinist, why it should occur at all, that is, it could be seen as the equivalent in biology of the economist Adam Smith's *invisible hand.* It is certainly the kind of pithy and memorable phrase that might appeal to someone who is setting out to educate school children. One has to conclude that Hunter deemed it unsuitable for some reason, as indeed Darwin himself ultimately did. [8]

4. 2. Authorial Stance / *Ethos*

A major element of authorial stance is what Aristotle in *Ars Rhetorica* calls *ethos*: the speaker's identity as it is constructed and/or deployed in discourse (Baumlin 1994; Cherry 1998). Ethos is particularly important in the context of CB because it is a fundamental tool in persuasion (Johnstone 2009: 10). By establishing his authorial stance, Hunter put himself in a position of authority over the reader-both educator and student. This manipulation of ethos, the successful presentation of himself as an authority, together with the misattribution of his own social Darwinist ideas to Darwin, leads to the reader to find more attractive and acceptable Hunter's version of evolution. The success of Hunter's authorial stance is attested by the fact that, as we point out in 3, neither the

[8] As can be seen from Table 3. 3., Darwin uses this term far less than most popular tradition would have it. Part of the reason for this is that it is a tautology, as has been pointed out by many scholars (e.g. Johnson 1993). In evolutionary terms, there can be no *a priori* definition for "fittest" and it can only be defined in terms of who survives.

prosecution nor the defence in the Scopes Trial questioned the fidelity of CB towards Darwin.

Other than being the author of this CB, very little information on George William Hunter is available. His dedication at the beginning of the work (which we reproduce as Figure 3. 1), is however revealing as it confirms that he was a high school biology teacher and also displays the two basic elements of *ethos*:

Dedicated
TO MY
FELLOW TEACHERS
OF THE DEPARTMENT OF BIOLOGY
IN THE DE WITT CLINTON HIGH SCHOOL
WHOSE CAPABLE, EARNEST, UNSELFISH
AND INSPIRING AID HAS MADE
THIS BOOK POSSIBLE

Figure 3. 1. Dedication to *Civic Biology*.

In this short passage, two features are in evidence, the first of which being the personal deictic device referring directly to speaker ("to *my* fellow teachers") and attitudinal epithets ("inspiring"). Pronouns (1st and 2nd) relating to addressor and addressee operate at the level of the interpersonal function of language. In the genre of textbook, they also have a directive function, given that the addressor has total command of the discourse and the addressee cannot talk back. Attitudinal epithets are a form of evaluation, establishing the author as a judge and thus placing him in a position of authority. In the following two subsections, we shall look at each of these features in turn.

4.2.1. Personal Deictic Devices

To analyse the frequency of personal pronouns in *Civic Biology* in Table 3.6, we compare the relative figures for CB and Darwin. Here as in Table 3.4, the figures are weighted to reflect the size of the respective sections of the corpus.

	ethos	↔	*pathos*
	I etc	*we* etc	*you* etc
Civic Biology	36. 98	212. 18	12. 58
Darwin Evolution	54. 26	47. 7	0. 23

Table 3.6. Personal pronouns in *Civic Biology* compared to Darwin.

As shown at the head of each column, *I*, *me*, *my*, *mine*, *myself* are orientated towards ethos, *you* etc. towards pathos with *we* etc. lying at the centre of a continuum between the two. [9] From the figures, it transpires that Darwin uses *I* etc. much more than Hunter but the latter uses *we* etc. and *you* etc. more. So there is a greater appeal in CB to *pathos* (appeals to characteristics of the audience - Johnstone 2009: 10) than there is in Darwin. So far the picture is typical. As a scientist proposing new theories, and often reporting in first person singular experiments carried out and observations in the field, it is not surprising that Darwin should use the first person often. Hunter, by contrast, is a textbook writer reporting the research of others and trying to incite interest in the reader in it. A greater frequency of the second person would therefore also seem natural. The data for *we* etc. is interesting, as it is notable that it is by a substantial margin the most common category of personal deictic device in CB (and a close second in Darwin) where the data suggest that *we* etc. are used as a convenient middle ground between overt ethos or pathos.

When one looks at the different sections of CB, the picture changes (Table 3.7.):

	ethos	↔	*pathos*
	I etc.	*we* etc.	*you* etc.
Civic Biology Foreword	18. 09	49. 75	0
Civic Biology minus evolution	3. 61	74. 12	7. 48
Civic Biology evolution	15. 28	88. 31	5. 09

Table 3.7. Personal pronouns in different sections of *Civic Biology*.

Table 3.7. shows that there are considerable differences between the various sections of CB. *I* etc. are used most in the foreword (where Hunter obviously follows convention and uses that forum to establish his ethos, in particular with the teacher / educator who has decided to adopt the book as a text). *We* etc. are most frequent in the sections on evolution and *you* etc. (pathos) are used least of all in the section dealing with evolution. Comparing the sections of the main text (excluding the foreword), there are noticeable differences between those that deal with evolution and those

[9] Within the category of *we* etc., one can distinguish between *exclusive* and *inclusive* according to whether the personal deictic excludes or includes the addressee as one of the referents, the former being closer to ethos, the latter to pathos.

that do not. In the former, there is a notably higher degree of ethos (15. 28 vs. 3, 61) and, in the latter, a marginally higher degree of pathos.

4.2.2. Evaluation and Stance

Expressions manifesting evaluation in some form or other are an element in the discourse related directly to authorial stance because, by using them, the addressor increases her position placing herself in a position to judge others. Hunston and Thompson (2000: 6-13) identify three functions of evaluation:

- To express opinion of addressor *vis à vis* the propositions expressed.
- To manipulate addressee's attitude *vis à vis* these propositions.
- To organize discourse (e.g. marking boundaries; highlighting).

In CB the main instruments of evaluation are *altitudinal epithets* and so-called *stance adverbs* (Biber *et al.* 1999) in the subsections below we shall look at each in turn.

4.2.2.1. Attitudinal Epithets

Attitudinal epithets are adjectives expressing the subjective attitude of the speaker towards something. They are a manifestation of the interpersonal function of language (see Halliday 2004). To measure the frequency of these, we compiled a representative list of common attitudinal epithets applicable to scientific ideas and their proponents and calculated the frequency of these in the various texts. In Table 3.8., we show the attitudinal epithets that we searched for, distinguishing between positive and negative attitudinal epithets (respectively PAEs and NAEs).

In Table 3.9., we show the figures (weighted as in Table 3.4.) for the frequencies for attitudinal epithets (positive and negative) comparing both CB with Darwin, and the various sections of CB.

Positive Attitudinal Epithets		Negative Attitudinal Epithets	
able	laudable	absurd	shallow
adept	learned	bigoted	simple
admirable	marvelous/marvellous	commonplace	stupid
adroit	masterly	crass	superficial
attractive	meritable	dull	trivial
capable	meritorious	fatuous	ungifted
commendable	noteworthy	foolish	unintellectual
copacetic	our/my fellow	heretical	unintelligent
deserving	perfect	ignorant	unlettered
earnest	praiseworthy	illiterate	unschooled
egregious	proficient	improbable	vacuous
esteemed	skilled	inane	
excellent	superior	incredible	
exceptional	terrific	insignificant	
expert	tremendous	intolerant	
exquisite	unselfish	irrational	
fine	valuable	narrow-minded	
good	wise	obtuse	
great	wonderful	pedantic	
greatest	worthy	prejudiced	
inspiring		puerile	

Table 3.8. List of attitudinal epithets related to scientific ideas and their proponents searched for in corpus.

	PAE	NAE
Civic Biology	50. 29	5. 37
Darwin Evolution	14. 3	1. 18
Civic Biology Foreword	22. 61	4. 52
Civic Biology minus evolution	11. 54	0
Civic Biology evolution	16. 13	0. 85

Table 3.9. Attitudinal epithets in corpus.

As can be seen (first two rows) PAEs are by far more common in CB than in Darwin. In this way, Hunter increases his own ethos, something perhaps more necessary for an unknown schoolteacher (see 4. 2.) than for an eminent scientist, by setting himself up as a figure qualified enough to praise people and opinions etc.

Looking at the individual sections of CB, as divided by us, (see bottom three rows) it can be seen that as with the personal deictic devices, Hunter imposes himself most in the foreword but significantly more in the sections on evolution than in those not on evolution.

Regarding the NAEs, it can be seen that in every case these are fewer than PAEs for both Darwin and CB. This reflects a general reluctance to being too confrontational. Hunter uses NAEs proportionately more often than Darwin thereby similarly appealing to ethos by establishing himself as someone able to criticise and pass negative judgment.

The figures for the various sections of CB also mirror the pattern for personal deictic devices. Attitudinal epithets of both kinds are used more in the foreword and less in the sections on evolution and in those dealing with matters not relating to evolution, a trend marked more with NAEs than PAEs. The general figures for attitudinal epithets show that it is on sections on evolution where Hunter is prepared to be marginally more contentious – something he does not attempt in other sections. This confirms independently the observations made in 4.1. regarding ideological stance.

4.2.2.2. Stance Adverbials

Another manifestation of evaluation comes in the form of what Biber *et al.* (1999) call *stance adverbials*, which can be subdivided into various kinds; the three main being *epistemic* (belief, knowledge), *attitude* and *style*. These we summarise in Table 3.10. with some examples of each kind (Biber *et al.* provide numerous examples in the form of various word classes and expressions).

The choice of names for some of the subcategories (i.e.evaluation) may cause some confusion in our discussion here because it is similar to terms that we have already used in other contexts here. The categories of epistemic and attitude relate most closely to evaluation as outlined in 4. 2. 2. although we would argue that doubt and certainty are better treated as two separate categories since they create potentially opposite types of authorial stance (a lack vs. a surfeit of authority). As all the subcategories of epistemic could be each seen in their own way as evaluative, *attitude*, *evaluation*, could, to avoid confusion, be dubbed (author's) reaction.

EPISTEMIC	Doubt and Certainly	*certainly; of course; I think.*
	Actuality / Reality	*actually; in fact; as a matter of fact.*
	Source /Evidence	*apparently; according to.*
	Limitation	*generally; on the whole*
	Viewpoint /Perspective	*in my opinion*
	Imprecision	*about; roughly; so to speak*
ATTITUDE	Evaluation	*amazingly; to my surprise*
STYLE		*confidentially; quite simply.*

Table 3.10. Stance adverbials (Biber *et al.* 1999 adapted).

In Table 3.11., we list the figures (weighted as in Table 3.4.) for the frequenciesof stance adverbs comparing both CB with Darwin, and the various sections of CB. The one change that we make is to distinguish between epistemic doubt and certainty:

	EPISTEMIC							ATITUDE	TOTAL
	C	D	AR	SE	L	VP	I	R	
Civic Biology	2. 06	0. 45	0. 45	0	0. 63	0	5. 52	0	9. 11
Darwin Evolution	3. 17	2. 9	0. 4	0. 02	0. 74	0. 01	2. 34	0. 01	9. 58
Civic Biology Foreword	0	0	0	0	0	0	0	0	0
Civic Biology minus evolution	0. 36	0. 45	0. 45	0	0. 63	0	2. 98	0	4. 87
Civic Biology evolution	1. 7	0	0	0	0	0	2. 55	0	4. 25

Table 3.11. Stance adverbials in corpus.

Looking firstly at the totals for all kinds of stance adverbials (last column), it can be seen that the figures for CB and Darwin are similar (9. 11 vs 9. 58) even though there are many differences between the subcategories. For Darwin, epistemics relating to doubt (D) and certainty (C) are markedly more prevalent than in CB, however, in both, certainty adverbials are more frequent than those expressing doubt although in Darwin the balance between the two is finer. In the other categories, the figures for AR (Actuality / Reality), SE (Source / Evidence), L (Limitation), VP (Viewpoint / Perspective) and R (Reaction) are very close. In CB, adverbs of imprecision (I) are proportionately almost twice more common than in Darwin. This last fact is explainable by the fact that Darwin, reporting his own and others' observations and experiments, needs to be more precise than Hunter who, writing for high school children, can content himself with giving more general data.

Examining the different sections of CB, it is apparent that while no stance adverbials at all are found in the foreword, the figure for the sections dealing with matters related to evolution and those dealing with other matters are very similar (4. 87 vs 4. 25). However, it can be seen that for CB minus evolution there are low figures for four of the eight categories (C, D, AR and L) with one higher score (over 50% of the total) for I. For CB evolution, by contrast, only two of the categories (C and I) occur but at a higher frequency for each. Certainty is thus more prevalent in sections on evolution and doubt entirely absent. It is here thus that Hunter's authorial stance as expert is emphasised. He voices relatively fewer doubts, ironically, on one of the more contentious areas of biology at the time of his writing (see 4. 1.). This may be the result either of his enthusiasm for social Darwinism or of a lack of confidence in his own grasp of the concepts involved, which he hopes to disguise with ostensible confidence. The latter explanation seems less likely seeing that CB was published one year before Hunt Morgan's seminal works (1915, 1916) made Mendel's work known to a wider audience, and Hunter shows himself to be well read in that he can already discuss him (see Table 3. 4.).

4.2.2.3. Modal Verbs as Markers of Stance

A final way that authorial stance can be analysed in CB is in the use of modal adverbs, epistemic and deontic. The former are of course closely related to epistemic adverbs (see section 4. 2. 2. 2.) and can be divided into certainty (e. eg. *will*, *should*, *must*, *can't*), and doubt (e.g. *may, might, could*). The latter are directives and, by using them, the addressor puts herself in an evident position of authority over the addressee. Due to the sheer number of such adverbs found in a corpus as large as ours, and due to the fact that a simple word search cannot distinguish between various uses of the same form (e.g. must as an epistemic as in "it must be true" and as a deontic as in "you must believe this") we selected a random sample of

50 using randomizer software[10]and estimated from this the proportions for the various uses. [11]

In Table 3.12. we give the results of this analysis, weighted as in Table 3.4.:

	Epistemic		Deontic
	Certainty	Doubt	
Civic Biology	27. 39	6. 41	26. 46
Darwin Evolution	34. 97	6. 09	14. 04
Civic Biology Foreword	18. 09	4. 52	81. 41
Civic Biology minus evolution	31. 49	2. 89	27. 82
Civic Biology evolution	3227	0. 85	14. 43

Table 3.12. Epistemic (certainty and doubt) and deontic modal verbs in corpus.

What is most striking is that, although epistemic modal verbs and epistemic adverbs must be seen as related, the figures on Table 3. 12. are markedly different from those on Table 3. 11. Darwin uses marginally more epistemic certainty modals than does Hunter in CB (as he does with epistemic certainty adverbs) but the figures for epistemic doubt modals are very similar.

Looking at the different sections of CB, it can be seen that the sections dealing with evolution contain slightly more epistemic certainty modals than those which do not. With epistemic certainty adverbs this difference

[10]Cyber-D's *list randomizer 1. 0*. This randomises by selecting items from a list only once, as is traditional; some statisticians prefer randomizers that allow items to be picked repeatedly.

[11] This method can be illustrated thus: together, the forms *must*, *will*, etc. were found 10, 277 times in Darwin. Taking a sample of 50 cases, we found them used as epistemic certainty modals 31 times (62%), which allowed us to estimate that in the whole corpus it would occur 6371. 74 times (i.e. 62% of 10, 8277). Assuming the worse-case percentage for accuracy (50%), the so-called confidence interval (i.e. margin of error) is ±13. 82. We can therefore reasonably expect the actual figure to be found somewhere between 6357. 92 and 6385. 56. The figure of 6371. 74 is then weighted as on Table 3. 4. to take into account the size of the corpus to give 34. 97.

is more marked. Of the three, the foreword shows the fewest epistemic certainty modals, but it contains no epistemic doubt adverbs at all.

Turning to deontic modals, an area we have not looked at yet, CB contains almost twice as many as does Darwin thus displaying an author who asserts his authority over the reader. Obviously, here the fact that Hunter was a high schoolteacher and the readers mainly students, may play a part. Darwin would seem to treat his readers more as his equals. In the different sections of CB, they are much more prevalent in the foreword than in either of the other two sections. In the section dealing with non-evolution related matters, they are almost twice as frequent as in those dealing with evolution. It would seem then that, contrary to the signals sent out in the other linguistic manifestations of stance that we have examined in Section 4., as far as deontics are concerned, he is less authoritarian in the sections of CB on evolution.

Hunter may appear more certain and less in doubt although he is less directive towards the reader. This means that he is attempting to convince the reader through the strength of his convictions not through his own stature, which is established in other sections. This use of a different strategy in the sections of CB dealing with evolution shows that Hunter does indeed single out the topic for a special means of presentation and may feel the need to reserve for it a distinct strategy in his exposition of what we can see is really his take on the subject despite his citations of Darwin.

5. Conclusions

What is striking in this study is not so much that Darwin is misrepresented--he being done so in a way fairly typical of the time--but the stance that Hunter takes in doing so.

It has been shown that it is precisely in those sections of *Civic Biology* that deal with evolution and related matters that Hunter interposes himself most, creating a more discernible and assertive authorial presence than in other sections of his book, relying more heavily on ethos but also on pathos in appeals to readers. It is also revealing that he expresses greatest certainty in sections dealing with evolution and doubt only in sections relating to non-evolution matters. In Darwin's original works, it can be seen that though ethos is present, it is less frequent and, in particular, he expresses doubt more often than certainty.

There is a degree of inconsistency in the aspects of evolution that Hunter cites: he refers to Mendel more than Galton, and this latter's eugenics would seem closer to his position and curiously he makes no mention of Spencer's well-known expression "survival of the fittest" although in his section on evolution, he speaks of "survival of the race".

As the full title of the work suggests (*Civic Biology Presented in Problems*), Hunter aspires to an inductive approach: something he makes explicit in the foreword; a desire that we would associate with interactive popularisation. It is ironic then that his domineering stance on evolution in particular means in effect that the students have little chance to learn about it inductively and are instead fed a series of prepackaged, sometimes radical, interpretations of Darwin and other relevant scientists' ideas.

Hunter's *Civic Biology* serves then as a warning of the way that popularisation and the processes that it involves can lead not only to the enrichment and dissemination of ideas which we would associate with interactive popularisation but to complex ideas being simplified and appropriated for largely ideological as opposed to purely scientific reasons.

CHAPTER FOUR

TALKING OF SCIENCE AGAIN AND AGAIN

GABRIELLA DI MARTINO

Throughout the entire history of science the need to popularise scientific results has been an important issue. Communicating science to the public nowadays is part of the general education process and it seems to be essential for people to share the results of scientific discoveries. Particularly today, due to the advance of science, which is faster than ever and has a wider impact on the general public, scientists and governments have to face questions which imply serious moral aspects and a great responsibility on how and what to communicate about scientific research to the general public. For instance, should the public have been informed about the potential catastrophic results of a nuclear explosion in 1949? (cf. John S. Rigden, in Teichmann et al. 2004).

The popularisation of science has long been considered a sort of public relations aiming to get the public's support. In the mid-16th century scholars used to discuss their scientific experiments in the so called coffee-houses which developed in London. They were a place for scholars to discuss science being a good example of how science moved from official institutions to public domains. The planned dissemination of science in general, and physics in particular, has been an important issue for the English scientific community since the end of the 17th century. It began to play an increasingly important role, carrying also elements of entertainment. Many physicists published "chatty books", pamphlets written in a light, familiar, conversational style, in order to reach, instruct and entertain as broad an audience as possible and also to increase their profits.

The prevailing belief was that orators and writers should write differently "from the vulgar maner and custome of writing or speaking" (Peacham 1593: 1). But very soon the Puritans began to assert that the common, simple, everyday language was beautiful and a plain style in preaching and simple words were considered more suitable to reach the understanding of most of the unlearned people who attended church on Sundays. John Wilkins declared that: "The principal scope of a divine Orator should be to teach clearly, convince strongly, and Perswade powerfully" (Wilkins 1653: 6).

The ornate sermon style was condemned as profane and one of the greatest Puritan preachers of the century, Richard Baxter, wrote:

> […] matter is first to be regarded as being of more Concernment than the Manner. […] It is commonly Empty Ignorant men that want the Matter and Substance of True Learning, that are over Curious and Sollicitous about Words and Ornaments […] all our Teaching must be as Plain and Evident as we can make it […] He that would be understood, must speak to the Capacity of his Hearers, and make it his Business to make himself understood. Truth loves the Light, and is most Beautiful when most naked (Baxter 1665: IV-38).

The Puritan ideal of Puritan literary style is beautifully represented by John Bunyan with his conversational prose and colloquial familiar speech of common people. The foreign-sounding sentences, particularly the archaic, Latinized structures, favoured by the rhetoricians came to be censured by the Puritans who supported practical, useful studies, thus contributing to the growth of the scientific movement. Science became a dominant interest which attracted aristocrats, businessmen and people of all sorts.

In order to transmit scientific knowledge it was necessary and urgent to create an appropriate medium of communication. Scientists needed a kind of prose that would accurately and clearly express their aims and principles in order to make accessible the applications of scientific knowledge to lay people. The Royal Society, founded in 1660, worked to complete the shift from the old to the new style in prose.

The important changes in the 17th century attitude towards English prose were due to the Puritan movement and to the scientific movement. They both promoted plain prose and condemned the old rhetoric. The Royal Society in the well-known manifesto clearly stated that:

> In all Reports of Experiments to be brought into the Society, the matter of fact shall be barely stated, without any prefaces, apologies, or rhetorical flourishes. Eloquence ought to be banished out of civil societies, as a thing fatal to Peace and good Manners. […] a constant Resolution [*is*] to reject all the amplification, digressions, and swelling of style: to retourn back to the primitive purity, and shortness, when men deliver'd so many things, almost in an equal number of words. They have exacted from all their members, a close, naked, natural way of speaking; positive expressions; clear senses; a native easiness: bringing all things as near the Mathematical

> plainness, as they can: and preferring the language of Artizans, Countrymen, and Merchants, before that, of Wits, or Scholars (Sprat 1667: 113).

Thus the Royal Society demanded plainness, clearness and directness of expression and its dominant aim was to improve the English Language particularly for philosophic and scientific purposes in order "to improve Natural Knowledge". The use of English instead of Latin was seen as a tool of communication for the spreading of scientific knowledge among the lay people beyond the limited circle of specialists.

Quite a number of philosophers, statesmen and scientists produced English translations of their most influential works. Francis Bacon composed his theoretical book *Novum Organum* (1620) in Latin, whereas his practical treatise *The Advancement of Learning* (1605) was written in English. William Harvey, who discovered blood circulation in the human organism, wrote about his findings in Latin in 1628, but also used the English language in his lectures on anatomy at the Royal College of Physicians in London, "interspersed with Latin words and phrases and so make a somewhat hybrid impression" (Glaser 1995: 192). *Principia Matematica* (1687) by Isaac Newton is in Latin, but his book *Opticks* (1704) is in English. The publications by Robert Boyle, one of the founders of modern chemistry, and Robert Hook, who developed a theory of light and colour, both members of the Royal Society, show their preference for English in the world of science. Thomas Burnet's *Theory of the Earth* was written first in Latin (1684) and then in English (1690).

Joseph Glanvill (1636-1689), a scientist/philosopher, was raised in a strict Puritan family and was a member of the Puritan Church; he was educated at Oxford University and became a fellow of the Royal Society in 1665. His work "is not all in one piece; he is definitely a 'transitional', and serves to link the ages of Browne and Boyle" (Willey 1952: 74). In the course of his life, in fact, he moved from the old, mannered, rhetorical prose style of Milton to the new, plain prose of Locke, as can be seen from a representative selection from his early to his middle and later style, through his main works: *The Vanity of Dogmatizing*(1661), *Scepsis Scientifica or Confest Ignorance, the Way to Science*(1665) and *Essays on Several Important Subjects in Philosophy and Religion*(1676).

In *Vanity of Dogmatizing* (1661) Glanvill attacks scholasticism and defends religious toleration, freedom of thought, and the scientific method. He attacks the old manner of writing as "a huddle of words and terms insignificant" (1661: 150) and appreciates the scientific method because "'tis built upon clear and settled significations of the names, which admit of no ambiguity, or insignificant obscurity" (1661: 160). Nevertheless his theories are not fully applied in his practice, as shown in the following example where the first two sentences are long and rhetoric, "prolix and involved, with long, subordinate clauses set off in an antithetic structure." The style of the last three sentences, on the contrary, "differs remarkably from the previous ones in their concrete, balanced brevity […] short and pointed" (Gazi 1959: 24).

> […] And as the Primogenial light, which at first was diffused over the face of the unfashion'd Chaos, was afterwards by Divine appointment gathered into the Sun and Stars, and other lucid bodies, which shine with an underived luster: so those scatter'd perfections which are divided among the several cantons of created beings, were as it were constellated and summ'd up in this Epithome of the greater World, Man.
>
> […] For since the most despicable and disregarded pieces of decay'd nature, are so curiously wrought, and adorned with such eminent signatures of Divine wisdome, as speak it their Author, and that after a curse brought upon a disordered Universe; what think we was done unto him whom the King delighted to honour? And what was the portion of Heavens Favorite, when Omniscience it self sat in Councel to furnish him with all those accomplishments which his specifick capacity could contain?
>
> Man was never at odds with himself, till he was at odds with the commands of his Maker.
>
> There was no jarring or disharmony in the faculties, till sin untun'd them.
>
> He could no sooner say to one power go, but he went, nor to another do this, but he did it (*V*: 2-4). [1]

The language of his first works, highly artificial, full of exotic, extravagant words, was in fact, the obstacle for his joining the Royal Society. So Glanvill reduced *Vanity of Dogmatizing* and published it with a new title, *Scepsis Scientifica*. In 1665, the same year when his membership was proposed again, he became a member of the Royal Society and he gradually conformed his style to a more concise and synthetic prose.

In his Address to the Royal Society, prefixed to *Scepsis Scientifica*, he writes:

> For I must confess that way of writing to be less agreeable to my present relish and Genius; which is more gratified with manly sense, flowing in a natural and unaffected Eloquence, than in the musick and curiosity of fine Metaphors and dancing periods. To which measure of my present humour, I had endeavour'd to reduce the style of these papers (*SS*: XXV-XXVI). [2]

---

[1] From now on *Vanity of Dogmatizing* is shortened to *V*.

[2] From now on *Scepsis Scientifica* is shortened to *SS*.

Here is a selection of examples taken from the two works. The reference to the poet and the Latin words in *SS* have disappeared and the sentences are shorter and more coherent:

> Some things we are inclined to love, and we know not why: others we disesteem, and upon no better account then the Poet did Sabidius, Hoc tantum possum dicere, non amo te. Some faces at first sight we admire and dote on (*V:* 21-22).

> And we love and hate without a known causeof either. Some faces both of Persons and Things, we admire and dote on (*SS:* 89).

Some images have been replaced by others which sound more familiar:

> I say therefore, that if we do but consider this Fabrick with minds unpossest of an affected madness (*V:* 43).

> I say therefore, that if we do but consider this Fabrick with free and unpossest minds (*SS*: 32).

> [it] is as unknown to us, as our embryo-thoughts(*V:* 43).

> [it] is as unknown to us, as the thoughts of our cradles (*SS:* 33).

Easier, more concrete terms replace uncommon ones:

> abstrusities (*V*: 27); difficulties (*SS:* 124).
> extern (*V:* 10); other (*SS:* 4).
> dictamens (*V*: 103); imagination (*SS.* 75).
> quinquarticular debates (*V*: 102); controversies (*SS:* 75).
> genuine derivation (*V:* 125); branch (*SS:* 93).
> drossie vehicle (*V:* 240); sense and body (*SS*: 177).
> indicate (*V:* 42); show (*SS*: 32).
> animal perfection (*V:* 44); perfect animal (*SS*: 33).

Some elements of vagueness have been omitted:

> [it] is an assertion, which, some say, hath no commission from our Faculties (*V*: 44).

> [it] is an assertion, which hath no commission from our Faculties (*SS*: 33).

> if we seek for an account how the parts of these do cohere, we shall find the cause to be as latent, as the effect of easy discovery (*V*: 49).

> if we seek for an account how the parts of these do cohere, we shall find our selves lost in the enquiry (*SS*: 36).

Glanvill seems strongly convinced that scientific theories have to be expressed with clear statements, settled sentence forms and with a simple vocabulary: "We […] should […] be as plain as we can, both in our Doctrines and Expressions; […] speaking in the proper, natural, easie way, which is most profitable for the ignorant, and most acceptable with the wise." (Glanvill 1676a: 103-104)

> Eloquence affects dangerously also peace and good manners since the generality are not capable of much Theory[…]mysterious, notional Preaching hath put many conceited People upon medlling with what they can never well understand, and so hath fill'd them with Air, and Vanity, and made them proud, phantastical, and troublesome; disobedient to their Governours, and contemptuous to their betters (Glanvill 1678: 19-20).

He declares his absolute, complete approval of the new manner of writing and he admires Sprat's effort to suit scientific language to common English and about his *History of the Royal Society* Glanvill writes:

> For 'tis manly, and yet plain; natural, and yet not careless: The Epithets are genuine, the Words proper and familiar, the Periods smooth and of middle proportion: It is not broken with ends in Latin, not impertinent Quotations; nor made harsh by hard words, or needless terms of Art: not rendered intricate by long Parentheses, nor gaudy by flanting Metaphors; nor tedious by wide fetches and circumferences of speech, nor dark by too much curtness of Expression: 'Tis not loose and unjointed, rugged and uneven; but as polite and fast as Marble (Glanvill 1668: 84).

Nevertheless it is only in his later works that Glanvill shows a radical change and his writing is totally consistent with the 17th century attitude of clear and unambiguous expressions.

The work considered here for a comparison is "Against Confidence in Philosophy", the first chapter of *Essays on Several Important Subjects in Philosophy and Religion* (1676) which is the revision of *Vanity* and in the *Epistle Dedicatory* Glanvill himself declares: "They [the *Essays*] were some of them written several years ago, […] now they come abroad together (with some things that are new) reduced to such an Order, as is most agreeable to my present judgment" (*Essays* p. a2).

And here is an example in which the same idea is condensed in a shorter paragraph, there are no omissions of ideas, the alterations affect only the expression:

> For all things being linkt together by an uninterrupted chain of Causes; and every single Motion owing a dependence on such a Syndrome of prae-required motors: we can have no true knowledge of any, except we comprehended all, and could distinctly pry into the whole method of Casual Concatenation (*V*: 213-214).

> All things are linkt together; and every Motion depends upon many pre-required Motors; so that no one can be perfectly known singly". (*Essays*: 16).

Formal revisions are massive in the *Essays*, long sentences are rephrased into short, clear ones, the constructions sound more familiar, superfluous or irrelevant phrases are omitted and more specific, concrete terms are used, for instance talking of an experiment "a winter night" becomes "a frosty night"; "aqueous Crystal" is substituted by "Water", "plantal emissions" is transformed into "Herbs".

In the *Essays*, metaphors give way to simpler, less emotional expressions and less technical terms: "Glandula Pinealis" is transformed into "Brain", "corporeal exuviae" is substituted by the more understandable "material images", "Progenitours" becomes simply "those before us". More concrete vocabulary takes the place of abstract: "sun" instead of "aethereal Coal", "spirits" instead of "subtil emissaries", "crepusculous obscurity" gives way to the more familiar "obscure Dawn", "midnight compositions" is simply "dreams".

In the *Essays* also the learned quotations disappear almost completely making communication easier for a wider public of readers.

A highly specialized philosophical/scientific text, to be popularised, has to be, in a certain way *translated* and this is what Joseph Glanvill did in the course of his life. Through a process of easyfication and recontextualization he tried to use the language of "Artizans, Countrymen and Merchants before that, of Wits, or Scholars" (Sprat 1667: 113) and rendered his texts more intelligible and more attractive for his readers.

Practical works and popular texts were written for the uneducated majority to make scientific information available to a great number of people and this feature brought a very important development to the discipline of science.

One of the first books on scientific theory written in vernacular, not in Latin, and explicitly addressed to lay readers, and specifically to women, was *Entretiens sur la pluralité des mondes* by Bernard De Fontenelle, published in 1686. In the preface the author expresses his purposes:

> I ought to inform those who read this work, and who have some knowledge of physic, that I have not here attempted to instruct them, but only to divert

> them, by presenting them in a manner something more agreeable and pleasing, than that which they already know (*Preface*).

The first translation of De Fontenelle's book into English by another Glanvill, John, appeared in 1687, followed in 1688 by a translation by Aphra Behn with the title *A Discovery of New Worlds.* Aphra Behn had a strong interest in Natural Philosophy and she had a critical look at the philosophers for not sharing their knowledge. It was a challenge for her to translate scientific texts especially since philosophy was considered a "masculine" discipline, however, Behn's translation "demonstrated her ability to participate in and translate between elite natural philosophy, often written in Latin (the most 'masculine' language) and a general or female audience" (Goodfellow 1996: 229).

Aphra Behn, about her translation of De Fontenelle, says: "I thought an English woman might adventure to translate anything a French woman may be supposed to have spoken" (Goodfellow 1996: 230). The book, in a discursive style, is not meant for an élite, but everybody can enjoy it. It takes the form of a dialogue between an astronomer who explains the Copernican heliocentric system of the universe to a dubious marchioness: "the earth is a planet, which turns round on its axis. And also revolves round the Sun."The astronomer expresses his opinion "that each Star may very probably be a Sun to enlighten other Worlds[…] that the Moon is a world like the Earth, and that apparently she's inhabited". He sounds very tentative and adds "I will not, however, swear that this is true; but I hold it for true, because it gives me pleasure to believe it" (De Fontenelle, *The second evening*). The astronomer explains how the planets and stars in the solar system are held in their orbits by forces called "whirlings" due to the centrifugal force. They go on debating about possible life on the moon, on whether the other Planets are inhabited and on what the inhabitants are like and the astronomer, very cautiously, makes very clear that the inhabitants of another planet are another family, they are not sons of Adam!

In De Fontenelle's prose and Behn's translation each author's assertions on the cosmological principles are beautifully illustrated by means of metaphors. The everyday language is scattered with personal references such as: "I have already said…" or "I forgot to ask how…" which give a very conversational tone to the text.

The 18th century, characterized by a general curiosity about nature, saw an increasing interest in observation in matters of science. Public demonstrations of science among educated lay people were very popular and also profitable and the experiments became the wonders of the age! Science had suddenly become fashionable.

In England, the great scientists were in touch with the continental scholars and their discoveries were communicated to the scientific societies. At the beginning of the 18th century, *The Philosophical Transactions of the Royal Society*, published regularly, was the only scientific periodical. *Lexicon Technicum: or an Universal English Dictionary of Arts and Sciences* by John Harris (1704) was the first technical dictionary in English in which mathematics and physical sciences are described. The second quarter of the century saw a great outpouring of literature aiming at educating the European élite in the latest scientific breakthroughs. Popular works such as the *Encyclopedia of Sciences, Arts, and Trades* published in France, was then read all over Europe.

One of the best-selling books was published in Naples in 1737 with the title *Il Newtonianesimo per le donne ovvero dialoghi sopra la luce e i colori*. The author, Francesco Algarotti whose explicit model was De Fontenelle, through lively conversations between a chevalier and a marchioness, describes and discusses some of Newton's experiments on light and colours. The book is devoted to the promotion of the popularisation of natural, experimental philosophy, meant for a female audience, "for ladies, that sex that prefer to feel rather than to know". It was translated into English (*Newtonianism for Ladies: Dialogues on Light and Colours*) by Elizabeth Carter two years later. Algarotti who joined the Royal Society in 1736 wrote: "May the age of things and of universal learning come upon us, when knowledge may serve not to harden hearts, nor to nourish the minds of the few, nor to make us suffer upon ancient and outmoded speech but rather, where possible, to improve and adorn society" (Algarotti 1739: xi, cit. in Mazzotti: 2004: 141).

So, popularisation was a well-known concept in past centuries even if the word was registered in the OED only in 1797 as a verb "To make popular, a. to cause to be generally known and accepted, liked, or admired; b. to render democratic." It is registered in the OED in 1836 with the meaning it has today: "to present (a technical subject) in a popular form" and, some years later, in 1871, "engaged in popularising history or science".

The popularised version of a *technolet* or "metamorphosis of a technical text for the general audience" is a practice known as a "scientific translation for the laity" (Glaser 1995: 180).

The author of a popularised text must "translate" a highly specialized problem for the perspective specialist. More explanations of technical terms, more illustrating examples from everyday life and more background information make a technical text more intelligible for the lay person. (cf. Glaser 1995: 180)

In all parts of the world, scientists have always had the urgent need of debating their discoveries in the advancement of science. It has always been like this in the history of science even at the time of the Emperor Fredrick II of Swabian when scientists from all over the known world would gather in his Court and discuss their scientific research and disseminate their results.

CHAPTER FIVE

THE POPULARISATION OF DARWIN(ISM) IN THE TWENTIETH AND TWENTY-FIRST CENTURY BRITISH PRESS: DARWIN AT DOWN HOUSE

KIM GREGO

Q [Interviewer]:
Was there a serious risk of your theory being rejected by the scientists?

A [Charles Darwin]:
It was touch and go for the first few years. There were a lot of quite negative reviews and some important scientists came out against it. Fortunately Huxley managed to get his very positive review into *TheTimes*, and that helped. Over the next few years, the positive comments gradually began to outweigh the negative ones (Bowler 2011: 78).

1. Introduction and Aims. Of Banknotes and Caricatures

While *The Guardian* informs us that Charles Darwin's portrait is to be replaced with that of Jane Austen on £10 notes (Rankin 2013), his reputation for scientific primacy does not currently seem endangered, although non-evolutionists still exist and resist. [1]The fact that Darwin is currently featured on the £10 note represents well his iconic strength at the popular level, his portrait being one of the best known in British culture, together with Winston Churchill's (next in line to be replaced on the five-pound note) (Peachey 2013) and Alfred Hitchcock's, who "would [also] look great on a tenner" (Davidson 2012).

[1] "There has certainly been a rise in what we call fundamentalist religious belief, and this has quelled widespread reaction against both evolutionism and the selection theory" (Bowler 2011).

Darwin was indeed one of the first men of science whose likeness was made widely public. Many caricatures of him, while often aimed at ridiculing his ideas, contributed instead to popularising his very distinctive appearance. [2] The advent of photography completed the task by passing on his image to future generations (Smith 2006;Browne 2009).

Banknotes and caricatures both testify to the degree of popular fame that Darwin has enjoyed in the 154 years since the publication of his *Origin of Species* (1859) [3]. First popularised in the 19th century – a period which saw an unprecedented spread of mass printing and especially of newspapers – the press remains the main popularising vehicle of Darwin's science and person to this day.

2. Background and "State of the Art": Darwin (ism) Made Popular

The scientific revolution and popular shock that Darwin's theory caused when outlined in his 1859 *Origin* granted his ideas some immediate immense popularity and notoriety, which continues well into the 21st century. It is only to be expected, then, that the literature on their popularisation should be extensive and varied—the relatively few years that separate Darwin's time from the present only adding to the interest for the topic and the quantity of the material written about it. This material, which at first glance might look homogeneous, can actually be divided into works on the popularisation of a) evolutionism, b) Darwinism as a specific instance of evolutionism and c) Darwin's figure—meaning Darwin as a person, as a public figure, and the relationship between these two aspects[4]. In turn, the scientific approaches adopted in researching these three aspects also vary considerably, and so do the temporal and spatial coordinates, length, level of specialisation and intended audience for the works. Although not all or even most of them can be mentioned here, a few select studies are worth reporting.

Historically, the first and one of the greatest popularisers of Darwin (ism) was his colleague and friend the biologist Thomas H. Huxley. The description and review of Huxley's contribution is of course better suited

[2] Various studies have investigated the iconic aspect of Darwin's fame. Among them, one recent work dedicated entirely to caricatures of the scientist, is Browne (2001).

[3] Hereinafter, *Origin*.

[4] Following the classification introduced, hereinafter the term "Darwin (ism)" will be adopted, to indicate b) and c) when co-occurring.

to studies focusing on the nineteenth century. Suffice it to mention here the first review on *The Times* of *Origin* (1859) (Huxley 1859: 8-9), his own treaty *Evidence as to man's place in nature* (Huxley 1863), and *Darwiniana* (1893), a collection of articles and lectures that

> either treat of the ancient doctrine of Evolution, rehabilitated and placed upon a sound scientific foundation, since and in consequence of, the publication of the *Origin of Species*; or they attempt to meet the more weighty of the unsparing criticisms with which that great work was visited for several years after its appearance; or they record the impression left by the personality of Mr. Darwin on one who had the privilege and the happiness of enjoying his friendship for some thirty years; or they endeavour to sum up his work and indicate its enduring influence on the course of scientific thought (Huxley 1893: v).

Over a century later, Numbers and Stenhouse (1999) offer a multilayered cultural perspective inclusive of place, race, religion and gender as factors influencing the reception of Darwinism in particular. The case studies analysed in this volume quote a variety of written sources, including newspapers, which "span the period from the publication of the *Origin* to the 1930s, when Darwin's theory of natural selection finally captured the allegiance of the scientific community" (1999: 1), and "illustrate the importance of local social and religious arrangements in affecting responses to Darwinism" (Ibid.). Perhaps the most original aspect of this work is the attention paid to the so-called "new worlds" like the US, Canada, Australia and New Zealand. These countries responded very differently to evolutionism, even in the case of geographically near countries like the US and Canada or Australia and New Zealand:

> Institutional maturity also seems to have made some differences. In Canada and Australia, for example, where non evolutionists frequently continued to occupy scientific chairs established before 1859, evolution entered the universities relatively slowly. In New Zealand, in contrast, which did not establish a university until 1869, evolutionists often occupied scientific chairs from their establishment. That made it easier in principle for evolution to gain a foothold. However, on occasion concern about the vulnerability of youthful institutions led their leaders to shy away from involvement in possibly damaging Darwinian debates (Numbers and Stenhouse 1999: 1-2).

Although the focus of Kelly (1981) is very restricted in both area covered (Germany) and time (1860-1914), its philosophical/socio-political perspective is also of interest, as it discusses the debated notion of "social Darwinism". This would be "the theory that societies, classes, and races are subject to and a product of Darwinian laws of natural selection"[5]. Specifically, Kelly deals with "popular Darwinism" as a German version of the phenomenon, linked to Marxism and referring in particular to the German working class. The reference material includes various printed genres, but it only extends to the beginning of World War I, because

> Although some Darwinists were still popular in the 1920s, especially among workers, [...].Deeply rooted as it was in the peculiarities of pre-1914 culture, [...]. After World War I nobody felt threatened enough by Darwinism to bother to fight it. Left without a benighted enemy, popular Darwinism lost much of its raison d'être. [...] Increasingly, popular Darwinism became a historical curiosity (1981: 145).

As will be remarked in detail in paragraph 3, this chronological remark also seems very relevant for an analysis of the popularisation of Darwinism that lies outside the socio-political scope, like the present one.

Yet, a different perspective is proposed by Harbottle (2004), the one-off work of a lay preacher turned atheist, discussing—from a philosophical and religious viewpoint—individual issues related to Darwinism such as "reason", "design", "Bible infallibility", etc. The primary source materials analysed are Darwin's *Origin*(1859) and Thomas Huxley's *Man's Place in Nature* (1863). Other sources considered span the whole of western culture and civilisation, and reflect what is both this work's strength and weakness: its' all-embracing and moral (though not moralist) approach, reminiscent of the classical humanist rather than the contemporary hyper-specialized academic. The portrait of Darwin it presents is, for instance, itself a masterful instance of popular writing:

> Darwin, the gentle grandfatherly figure, is a popular image: charming, disarming, loveable, childlike – a modest, retiring invalid quietly ploughing his lone scientific furrow with no thought of worldly fame. There was, however, a different Darwin behind this benign façade—a calculating man of ruthless ambition who exploited almost all with whom he came in contact. While a student, for example, he imperiously expected lesser mortals to carry birds he had shot and made no allowance for his companion Herbert's lameness by expecting him to tramp painful miles catching insects for him (Harbottle 2004: ix).

Harbottle (2004) is a clear example of how one common issue among scholars writing on Darwin and popularisation is the temptation to

[5] Oxford English Dictionary 2013 online ed., s. v. "Social".

contribute to the popularising process itself by expressing personal views and drawing biased portraits of the scientist.

Getting closer to the specific interest of this study, i.e. Darwin (ism) in the press, but still focusing only on the 19th century, Horenstein (2009) is a paper worth mentioning because it reviews the "reaction to Charles Darwin's *Origin of Species* and his later works in the 'popular press' of the United States [...] between 1859, the year of the publication of the *Origin of the Species* and Darwin's death in 1882" (2009: 107). Much in line with this paper's own purposes, it is limited, though, to a compilation of a conspicuous and detailed list of quotations from newspapers, magazines and various publications, not followed—regrettably—by any real critical commentary.

Coming to the historical perspective, this sees two main scholars writing on the popularisation of science in general and of Darwinism in particular: Peter J. Bowler and Bernard V. Lightman. Peter Bowler, Emeritus of History of Science at Queen's University, Belfast, is the author of many publications on the topic, including one of Darwin's recent biographies (1996). Most deal with the fortunes suffered by Darwinism in history. In his 1983 review of the pro-and anti-Darwinist debate in the English-speaking world, based on reference material belonging to various genres (and in its update of 2005), Bowler only gets to the 1920s, because "by 1920 some geneticists were beginning to admit a limited role for selection by conceding that useful new characters were far more likely to spread than harmful ones" (1983: 213). Indeed, by the late 1940s, Darwinism as a scientific theory was no longer in danger: "At another conference held in Princeton in 1947 there was an almost embarrassing lack of discussion, so great was the success of the new Darwinian theory" (Ibid.). Bowler's greatest contribution to the popularisation of Darwin via the press, though, is his 2009 work, itself showing at least a partial popularising intent. As in all the above mentioned pieces of research, this book analyses the situation up to the early twentieth century, i.e. approximately the 1920s, but it draws an informed and accurate picture of the historical context of the (popular) press in that period. "By 1900", he writes, "there was almost universal literacy. This had allowed the emergence of mass-circulation newspapers and magazines such as the *Daily Mail* and *Tit-Bits*, scorned by the intellectuals as being written for shop boys with limited education and attention spans" (Bowler 2009: 81). However, the "Great War caused a massive interruption in the publishing trade. There were shortages of raw materials, and marketing was interrupted" (Bowler 2009: 133). As regards newspapers in particular:

> The press became increasingly dominated by "barons" such as Alfred Harmsworth, Lord Northcliffe, whose *Daily Mail* pioneered the techniques of sensationalism and the cult of the personality. As an imperialist, Harmsworth had some interest in applied science, but little of this found its way into the pages of his newspapers. Again, it was technology, especially the technology of speed, which could be used to capture the popular imagination (Bowler 2009: 186).

And, to report on specific newspapers,

> Although the *Times* had a science correspondent from the turn of the century, J. G. Crowther's work for the Manchester Guardian in the 1930s pioneered the profession in its modern form. Few working scientists could acquire the skill needed to produce copy of the restricted length (only a few hundred words) demanded by the daily press. E. Ray Lankester's "Science from an Easy Chair" articles for the *Telegraph* became justly famous and managed to smuggle in the occasional more serious comment. J. Arthur Thomson and W. P. Pycraft were the only other professional scientists to write regularly for the daily or weekly press in the interwar years, both specializing in natural history (Bowler 2009: 187).

Much as in today's world, "Publishers and authors thus had to strike the right balance between education and entertainment" (Bowler 2009: 81). In Chapter ten, Bowler continues the historical review of the popularising British press in the late eighteenth and early nineteenth century, with the interesting section "Science in the Newspapers" (2009: 196-202), and goes over the various writers for papers and magazines specializing in science at that time in the following section "The Science Correspondents" (2009: 202-209). Bowler, however, is founded on reference works, but does not make use of any specific corpus of newspaper material. Other works by Bowler tend towards popularisation, like the very recent and speculative *Darwin deleted: imagining a world without Darwin* (2013), or the successful Q&A format of *Darwin: off the record* (2011), in which the author relates an imaginary interview with Darwin, who thus "speaks" in the first person—a classic yet still effective rhetorical device.

Bernard V. Lightman, Professor of Cultural History of Science at York University, Canada, has also written extensively on science popularisation (Lightman 2007) and on Darwin (Lightman 2009 and 2010), but focusing on the Victorian period and favouring a culturalist historical perspective (for example, concentrating on gender and the visual, as in Lightman and Shteir 2006). Lightman (2010), as regards place and time, reports on "how evolutionary theory was popularised in Britain and the USA from 1860 to 1900" (Lightman 2010: 6). The approach, "In contrast with Peter Bowler, who uses categories based on the mechanism of evolution in his discussion of the positions of elite scientists in his book *The non-Darwinian revolution*" (Ibid.) is based on categories "related to the religious and metaphysical concerns of the popularisers" (Ibid.). Concerning the materials studied, it deals "only with those popularisers who accepted evolution as valid and who took it upon themselves to write books explaining the scientific basis of the theory and its larger intellectual implications" (Ibid.). No newspapers are considered.

Last but indeed the most relevant for this project's purposes is Ellegård's book on the reception of Darwinism by the general public (1958), since its author has "in the main relied on an investigation of the British newspaper and periodical press of the time" (1958: 7), and scrutinised "well over a hundred titles of periodicals" (1958: 8). Dated as it is, Ellegård's one-off work remains the leading study on the popularisation of Darwin (ism) *through* the British newspaper press. Its limits are several, and not only represented by its 55 years of age. To start with, the time span considered is 1859-1872 which, in the author's opinion, "is short enough to allow a fairly complete coverage of the discussion" (Ibid.). In fact, it can be argued that the period is short in absolute terms, and that there seems to be no rationale behind the selection of the "first dozen years after the publication of the *Origin of Species*" (1958: 7) except the author's own subjective choice. Where others (e.g. see Horenstein 2009 above) have chosen Darwin's death in 1882 as the end limit, in this case, for instance, the publication of the *Descent of Man* could have represented a reasonable selection criterion. However, this appeared in 1871, thus making 1872 an arguable date in which to end the review, since it includes part of the reactions to *The Descent of Man* but certainly not a chronologically thorough account of them. Secondly, the book reflects the opinion-based essayist approach of its time, apparent in the structure of the content, which includes chapters generically titled "The Argument of design", "Miracles", "The Bible", etc. The four main influences on the reception of Darwin (ism) identified in the book are also quite differing and all-embracing: politics, religion, philosophy of science and science itself, necessarily resulting in a generalist rather than a specific exposition. On the other hand, Ellegård (1958) presents with an innovative and invaluable feature that places it decades ahead in research at its publication: the argumentation is based on solid evidence in the form of a corpus of newspapers and periodicals from the time span considered. In

David L. Hull's[6] own words from his foreword to the 1990 edition, "Increasingly, historians are coming to recognize the need to base their conclusions on more solid foundations. Ellegård anticipated this trend by more than two decades" (Ellegård 1958/1990: 1). The corpus comprises material extracted by reviewing 115 periodicals and newspapers, of which 107 proved relevant. On the whole, "It has been calculated that 546 million copies of newspapers were sold in England and Wales in 1864; my newspaper sample covers c. 189 million, or about one third of this, and includes almost all the important London papers" (Ellegård 1958: 338). Although the corpus is not included in the book (and could not at the time have been provided on a separate digital support, either on-or off line), the two Appendixes—"Statistical analysis of the press reaction. Discussion of methods" and "The periodicals: classification, coverage of Darwinism, index of references"—are, with their numerous data systematised into tables and statistical graphs, a precious resource for anyone interested in a corpus-based analysis from any scientific perspective. Indeed, given the presence of the corpus, even within the arguable choice of period discussed above, Ellegård's classification of the popularising phenomenon into three sub-periods—1859-1863, 1864-1869 and 1870-1872—proves convincing. The first sub-period, starting with the publication of *Origin* in 1859 and finishing with that of "T. H. Huxley's *Man's place in nature* and above all Sir Charles Lyell's *Antiquity of man*" (1958: 28) in 1863, is very productive and represents most of the material in the corpus. The second sub-period

> was on the whole quieter on the Darwinian front, except in 1868, during which Darwin published Variation of animals and plants under domestication, in which was presented part of the evidence on which the discussion in the *Origin* had been built, and the theory of Pangenesis was launched (Ibid.).

The third sub-period is only two years and, thus, objectively short and disproportionate to the others, as argued earlier on. It obviously includes Darwin's own *Descent of man* (1871) and

> other substantial contributions to the Darwinian discussion during these years, notably A. R. Wallace's *Contributions to the theory of natural*

---

[6] Incidentally, Hull, Professor of Philosophy at Northwestern University until his death in 2010, is also the author of another relevant work on the subject from the 1970s: *Darwin and His Critics: The Reception of Darwin's Theory of Evolution by the Scientific Community* (1973). This, like other interesting studies falling within the perspective of philosophy, could not be described in this limited exposition.

> *selection*, and St. George Mivart's *Genesis of species*. Both these books were important in switching the centre of interest from the Evolution theory as such, which both writers advocated, to the Natural Selection theory, in which both found difficulties. Mivart, especially, may be said to have struck the note of public opinion during this period (Ibid.).

On the whole, Ellegård (1958) remains a milestone in both history of science and in corpus-based humanities studies in general. For the specific purposes of this paper, it is doubly relevant, as a) its collection of periodicals—newspapers in particular—represents a key starting point for the setting and the analysis proposed here; and b) its author's multi-disciplinary approach, although it often proves too generic and thus vague, is also extremely contemporary, and the stated nature of the work as a "contribution to what may be called the social history of idea" (Ibid.) very well links, through the adjective "social", to the Critical Discourse Analysis perspective followed here.

3. Aims: Darwin or Darwinism?

The present paper looks at how Darwin and his thought were popularised in the 20th-century British press, focusing on the specific case of *The Times*.

The research questions pursued in this investigation were:

- How was Darwin (ism) popularised in the reference corpus?
- Was the focus of the popularisation Darwin or Darwin's ideas?
- What conceptual and linguistic strategies of popularisation were employed?

The answers to the above questions have been pursued through a quality-based lexico-semantic analysis of a sample corpus of *Times* articles as described in section 4, with the results presented in section 6.

4. Material: The "Darwin in British Newspapers" Corpus

Charles Darwin himself "believed that the ultimate success of his theory of evolution by natural selection depended on its acceptance by the Victorian popular audience, not just scientists" (Lightman 2010: 6). The famous British naturalist, clearly, very well perceived the importance of winning over popular audiences, and even encouraged his followers to do so. His correspondence with T. H. Huxley, to name one, is famous for

Darwin's invitations to him to "spread a taste for the Natural Sciences"[7]. Heeding Darwin's own exhortation, then, the decision to devote a specific research project to science popularisation via the British newspapers appears like "natural selection".

The "Darwin in British Newspapers" (Dar. Bri. N.) Corpus is the provisional name given to the prospective result of an ongoing project, aimed at screening a selection of British newspapers (dailies) for instances of popularisation of Darwin (ism). Ideally, the corpus is supposed to span from the 1840s to the present day, receiving periodical updates as required and possible. The project was started by Giovanni Iamartino, Professor of English at the University of Milan, Italy, in 2012, and so far includes 44 British periodicals electronically scanned, which produced approximately 350 articles of varying relevance for the topic.

Two preliminary lines of investigation are currently being pursued within the project: one looking at the late nineteenth century and early twentieth century, and one focusing on the mid-to late twentieth and early twenty-first century. This draft categorisation can be visualised as follows:

a. 1840s-1920s: Darwin's affirmation, publication of *Origin*, decades following death, WWI;
b. 1920s-2010s: soon after WWI to the present day.

This is only one of many possible chronological divisions, aimed at obtaining two initial sub-groups of material. The decision to see the 1920s as a turning point is based, among other considerations, on the findings reported in Kelly (1981) and Bowler (1983) about Darwinism having reached, by that time, an established scientific status. Although "The need for scientific popularisation had probably never been greater than in the 1920s" (Kelly 1981: 145), starting from after WWI, Darwin's ideas—at least in academic / professional circles—turned into the mainstream view on evolution, and their difficulty in being accepted diminished. The contingencies of the war, with the surprising and often lethal applications of science to warfare in the form of technology, might also have contributed to raising the (negative) consideration of science among the population at large and bringing them closer to views of the world previously deemed "impossible". The material being collected for the Dar. Bri. N, of course, can and ought to be divided and analysed according to more and diverse criteria, depending on the specific research meant to be carried out on it.

[7] Letter from C. R. Darwin to T. H. Huxley, December 7, 1862.

4.1. The *Times* Corpus

This paper, in particular, uses the Dar. Bri. N material drawn from a single newspaper, *The Times*, from period b. (1920s-2010s). Founded in London in 1785 and having taken up its current name in 1788, *The Times* had a circulation of 394, 982 as of June 2013[8], which places it well in line with other quality newspapers nationally and worldwide, and at number 48 in the world in 2011[9]. Given the history and the relevance of the newspaper considered, the quantity of the articles about Darwin (ism) since the 1840s is conspicuous. Ellegård (1958) well describes *The Times*' reaction to the natural selection theory throughout the years, which may be summarised as follows.

Starting from 1859,

> The newspapers of Mid-Victorian Britain usually contained only brief literature sections. In the better-class papers fairly substantial reviews of the more important books used to be published. The *Times*, for instance, had on an average 100-150 reviews per year, of which many were very brief, while others extended to 5-6 columns. The popular papers, on the other hand, when they carried a literature section at all, mainly offered brief notices only, or short extracts or episodes from the works noticed. The selection of works reviewed was decidedly trashy [...] . Under the circumstances reviews of Darwin's book could only be expected in the more expensive newspapers. In fact, the *Times*, the *Morning Post* and the *Daily News* were alone among the morning dailies to review the *Origin* (Ellegård 1858: 25).

Of the dailies that did publish a review of the *Origin*,

> only the *Daily News* took a decided stand against Darwin [...] . The *Times* was favourable indeed: by an incredible stroke of luck, its review came to be entrusted to Huxley, who thus found a golden opportunity to present Darwin's work to its best advantage to a large and influential public. The *Times* was at that time not only by far the most important English newspaper, it had also the largest circulation of all. The other newspaper reviews were fair, and on the whole quite favourable (Ellegård 1858: 29).

In time, the attitude of *The Times* changes for the worse, taking a conservative turn: "Gradually the normal spirit of the paper asserted itself

---

8 "ABCs: National daily newspaper circulation June 2013, " guardian. co. uk (July 12, 2013).

9 The International Federation of Audit Bureaux of Circulations (IFABC), "Circulation data for national newspaper" (January 17, 2013).

in later pronouncements on Darwinian questions, and when the *Descent of Man* appeared, the *Times* review was violently hostile" (Ellegård 1858: 36).

By 1871, the *Times*' attitude toward natural selection was that, on the whole, if the organic part of humans might have been resulted from evolution, their soul remained a totally distinct affair (Ellegård 1858: 315). The readership of *The Times* in the 20th century has been amply discussed, for example by Bell (1991). Remarks on the contemporary (i.e.21st century) diastratic positioning of *The Times*, its current formats and circulation appear, for instance, in Vicentini, Grego and Russo (2013).

4.2. The Down House Sample Corpus

For the present analysis, the need was felt to further reduce the *Times* corpus to a smaller, sample corpus that could nonetheless be representative of a specific aspect of the popularisation of Darwin (ism) and be used to try out the methodology and research approach.

One interesting and recurring subject to pursue in the *Times* corpus is represented by Down House, the Darwin's family home in Kent that saw the writing of *Origin* as well as of his following treaties, and a number of experiments that the botanist conducted in the estate until his death, in his home, in 1882. Of the mansion, Darwin wrote to his sister the year he moved in: [10]

> My dear Catty
> You must have been surprised at not having heard sooner about the House.
> [...]
> I will give you in detail [...] my opinion on it. [...]
> House ugly, looks neither old nor new.
> [...] not too near or too far from other houses
> [...] [Village] Inhabitants very respectable.

Table 5.1. shows a concise chronology of Darwin's life before and at Down House, and the house's history following his death.

[10] Letter from C. R. Darwin to sister Catherine, July 24, 1842.

Year	Down House Chronology
1809	Charles Darwin born
1831-1836	Darwin's voyage on the Beagle
1842	Moved in from London on 24 September 1842 with wife and two children, had seven more in Downe
1859	On the Origin of Species published, written at Down
1882	Charles Darwin dies
1896	Wife Emma dies
1900-1906	Let to tenant
1907-1927	Girls' boarding school
1927	Bought by British Association for the Advancement of Science
1939-1944	Closed during WW2
1953	Donated to Royal College of Surgeons
1954	Became Grade I listed building
1962-1983	Sir Hedley Atkins lived at Down House as curator with wife
1996	Bought by English Heritage
1998	Down House reopens as a museum
2009	Darwin's centenary

Table 5.1. Down House chronology

Because of the key role that Down House had in Darwin's experiments[11], in his developing and putting into writing his evolutionary theories, and in his personal life as a husband and father to a large family, a corpus about the property was believed to have the potential to return data on both Darwin's figure and Darwin's science.

An *ad hoc* sample corpus for this research paper was thus built, according to the following selection criteria:

- articles should come from *The Times*, print or digital edition;
- articles should be from the period 1920-2013[12];
- articles should include the words DARWIN AND DOWN AND HOUSE.

[11] Of which those on worms are possibly the best known by the general public, cf. Charles Darwin, *The Formation of Vegetable Mould through the Action of Worms*, with Observations on their Habits, London: John Murray, 1881.

[12] End date: June 30, 2013.

Articles were retrieved using the text search engines of *The Times Digital Archive (1785-2006)* [13] and of *The Times* digital archives available at the British Library, London (1788 to date). The initial selection of texts was later scanned in person and qualitatively for content and relevance.

The selection returned a corpus of 41 articles as in Table 5. 2. below.

No.	Date	Title
1	1927. 09. 02	Darwin's house. Proposed purchase for the nation.
2	1927. 09. 02	Charles Darwin's home.
3	1927. 09. 05	The Estate Market. Land at Downe.
4	1929. 0. 25	Charles Darwin's home. A national memorial.
5	1929. 02. 25	British Association and Darwin's house
6	1930. 03. 18	Rating of Down House. British Association summoned
7	1933. 03. 13	Down House
8	1933. 03. 31	Darwin's home
9	1934. 03. 31	Charles Darwin's home
10	1935. 04. 01	Down House
11	1935. 11. 26	Down House
12	1937. 09. 08	Downe – A dangerous honour
13	1942. 11. 03	MSS. of Charles Darwin. Gifts to Cambridge and Down House
14	1962. 03. 23	Darwin orchid site bought by Trust
15	1962. 03. 26	Picture Gallery, Downe Bank, Kent
16	1963. 02. 01	Repair grant for home of Charles Darwin
17	1968. 04. 06	Historic homes
18	1973. 06. 08	Charles Darwin Lived Here. BBC2
19	1985. 11. 16	Home truths on Darwin, the family man
20	1995. 03. 20	Darwin's haul
21	1998. 01. 28	£2m restoration of Darwin's home
22	1998. 03. 05	Darwin House
23	1998. 03. 28	Survival of the original naturalist
24	1998. 10. 13	Rare Darwin photograph on sale
25	2000. 04. 19	Darwin stones natural choice for museum
26	2008. 11. 12	The original of the species—Arts
27	2008. 12. 26	Darwin and Discovery—Arguments of 2009: the theories of a great scientist
28	2009. 01. 03	Britain's Galápagos offers insight into evolutionary ideas
29	2009. 01. 22	'It always seemed clear that we were related to monkeys'—On the 200th anniversary of Charles Darwin's birth, Sir David Attenborough tells Damian Whitworth why the great scientist 'had the answer to most things'

[13] *The Times Digital Archive (1785-2006).* London: Natkins.

No.	Date	Title
30	2009. 02. 07	Darwin's life in measured verse—poetry
31	2009. 02. 10	Discover why even the call of nature didn't stop Darwin
32	2009. 02. 12	I'd love to tell Darwin he was right all along—Was the evolution theorist just a beetle-collector who got lucky? No, says Armand Leroi. He was an icon whose discoveries echo through every branch of 21st-century science
33	2009. 02. 12	In search of Darwin
34	2009. 02. 13	Going out... Staying in
35	2009. 03. 23*	Life is short but this delightful poem lives on—After 150 years, *The Rubaiyat* of Omar Khayyam still has an uncannily modern moral—enjoy yourself while you can
36	2009. 11. 24	Hunt for Darwin notes
37	2010. 05. 23	2 for 1 entry at English Heritage
38	2010. 10. 23	Great half-term days out
39	2011. 07. 20	Going out... Staying in
40	2012. 02. 15	Going out... Staying in
41	2012. 06. 30	30 great historic days out—Free family ticket to English Heritage sites worth up to £42

Table 5.2. Corpus of articles.

5. Methodology: From Social History to Critical Discourse Analysis

The analysis presented below is a quality-based, lexico-semantic investigation, which bases its approach on Critical Discourse Analysis (CDA) for its descriptive, ethical and ideological stance, with a special focus on promotional discourse (as in Fairclough 1992, 1995a, 1995b, 2003 and 2006; van Dijk 1988 and 2008; Kress and Hodge 1979; Chouliaraki and Fairclough 1999; Wodak 2013).

For their relevance in prescriptive, specialized settings, views from Critical Genre Analysis (CGA) were also adopted, as in Bhatia (2004 and 2008). In particular, as regards the popularisation of science, reference was made to recent works by Gotti (2013), Grego (2013) and Vicentini, Grego and Russo (2013).

Although WordSmithTools 5.0 has initially been used to draw frequency lists and concordances of terms, this sample corpus's reduced size does not actually justify the use of concordancers, which would however come in useful, for the enhanced speed and power of analysis they provide when working with larger corpora.

The definition and model of popularisation adopted refer to Grego (2013: 151-154) for the study of English for Specific Purposes genres, within a wider view of the discourse of popularisation as defined by lack of discussion, non-specialist target audiences and informative purpose (Gotti 2013: 9-10).

In brief, the attempt has been made to investigate the role of the 20th and 21st century press as a fundamental medium in science popularisation, following a CDA perspective attentive to ideology, society, genres as communicative vehicles in professional settings.

6. Analysis: A Man and His House

6.1. Lexico-Semantic Areas

The lexico-semantic reading of the texts in the Down House sample corpus resulted in the identification of three main semantic areas:

1. the *person* of Darwin—man, scientist, symbol;
2. the *science* of Darwin—scientific dissemination (?);
3. the *economy* of Darwin ('s house)—monetary aspects linked to Down House.

The categorisation is perhaps clearer as seen in Table 5.3., where P stands for point 1., S for point 2. and E for point 3.

No.	Date	Title	Lexico-semantic area
1	1927. 09. 02	Darwin's house. Proposed purchase for the nation.	E
2	1927. 09. 02	Charles Darwin's home.	E
3	1927. 09. 05	The Estate Market. Land at Downe.	E
4	1929. 02. 25	Charles Darwin's home. A national memorial.	P
5	1929. 02. 25	British Association and Darwin's house	E
6	1930. 03. 18	Rating of Down House. British Association summoned	E
7	1933. 03. 13	Down House	P
8	1933. 03. 31	Darwin's home	P, S, E
9	1934. 03. 31	Charles Darwin's home	P, S, E
10	1935. 04. 01	Down House	P, S, E
11	1935. 11. 26	Down House	P

No.	Date	Title	Lexico-semantic area
12	1937. 09. 08	Downe – A dangerous honour	E
13	1942. 11. 03	MSS. of Charles Darwin. Gifts to Cambridge and Down House	S
14	1962. 03. 23	Darwin orchid site bought by Trust	S
15	1962. 03. 26	Picture Gallery, Downe Bank, Kent	S
16	1963. 02. 01	Repair grant for home of Charles Darwin	E
17	1968. 04. 06	Historic homes	E
18	1973. 06. 08	Charles Darwin Lived Here. BBC2	P
19	1985. 11. 16	Home truths on Darwin, the family man	P
20	1995. 03. 20	Darwin's haul	E
21	1998. 01. 28	£2m restoration of Darwin's home	E
22	1998. 03. 05	Darwin House	E
23	1998. 03. 28	Survival of the original naturalist	P
24	1998. 10. 13	Rare Darwin photograph on sale	E
25	2000. 04. 19	Darwin stones natural choice for museum	S, E
26	2008. 11. 12	The original of the species—Arts	S, P
27	2008. 12. 26	Darwin and Discovery—Arguments of 2009: the theories of a great scientist	S
28	2009. 01. 03	Britain's Galápagos offers insight into evolutionary ideas	S
29	2009. 01. 22	'It always seemed clear that we were related to monkeys'—On the 200th anniversary of Charles Darwin's birth, Sir David Attenborough tells Damian Whitworth why the great scientist 'had the answer to most things'	S
30	2009. 02. 07	Darwin's life in measured verse—poetry	P
31	2009. 02. 10	Discover why even the call of nature didn't stop Darwin	P, S
32	2009. 02. 12	I'd love to tell Darwin he was right all along—Was the evolution theorist just a beetle-collector who got lucky? No, says Armand Leroi. He was an icon whose discoveries echo through every branch of 21st-century science	P, S
33	2009. 02. 12	In search of Darwin	P
34	2009. 02. 13	Going out... Staying in	P
35	2009. 03. 23	Life is short but this delightful poem lives on—After 150 years, *The Rubaiyat* of Omar Khayyam still has an uncannily modern moral—enjoy yourself while you can	(P)

No.	Date	Title	Lexico-semantic area
36	2009. 11. 24	Hunt for Darwin notes	S, P
37	2010. 05. 23	2 for 1 entry at English Heritage	E
38	2010. 10. 23	Great half-term days out	E
39	2011. 07. 20	Going out... Staying in	E, P
40	2012. 02. 15	Going out... Staying in	S
41	2012. 06. 30	30 great historic days out—Free family ticket to English Heritage sites worth up to £42	E, P, S

Table 5.3. Lexico-semantic areas

As can be inferred from Table 5.3, not all the articles can be labelled univocally as belonging to just one category. Six of them (25, 26, 31, 32, 36 and 39) are best described as belonging to two categories, and four (8, 9, 10 and 41) as falling into all three of them. On the whole, 19 articles can be said to include lexicon pertaining to Person, 16 to Science and 19 to Economy. Considering the relevance of Darwin in late-modern and contemporary science, the quite even balance between the categories might appear surprising. A historical reading of the corpus can perhaps shed light on the reasons why. The first documents on Down House are from the late 1920s, and they all deal with the economic aspect of the estate. This is only to be expected, when looking at the chronology of the house (Table 5. 1.): up until 1927, Down House was in use as a boarding school, so its fate only came to prominence in that year. That is when, some 45 years from the death of the scientist, Britain started to interrogate itself about the value—scientific, historical, economical—of the property. The trend continues during all the 1930s, the question resolved by the purchase of the house by the British Association for the Advancement of Science, but constantly needing funds in a period of economic hardship with little private or public money to spend. In this decade, three consecutive articles appear which contain lexicon from all three categories (8, 9 and 10): these must be read as appeals for sponsorships for the mansion, in which the person of Darwin and his scientific heritage are called upon to pursue the cause. In the 1940s, Britain, like the rest of the world, was preoccupied with more pressing international matters. There is only one article from 1942 in the corpus (11): a few lines reporting a donation of some of Darwin's personal belongings to Down House, which remained closed throughout WWII. In the 1950s, the estate was donated to the Royal College of Surgeons, and became a Grade I listed building, but still suffered heavy financial hardship. No significant articles appeared about it in that decade in *The Times*. The 1960s were the years of the economic boom and swinging London. In 1962, Down House was bought by the Kent Naturalists' Trust, and for some 21 years it was inhabited again, by Sir Hedley Atkins, former President of the Royal College of Surgeons, who lived there as a curator with his wife—and with this the chronology enters the 1980s. In the intervening decades, the few articles altogether published about Darwin's property belonged to various categories. In the 1990s the economic factor rises again, continuing into the early 2000s. Then nothing is heard again until 2006, the glorious year in which Down House was finally bought by English Heritage: from that moment up until and including 2009, the articles about the estate are almost all concerned with the scientific aspect—the property being eventually free from debt (or not openly venting that preoccupation) and journalists free to concentrate on popularising science alongside the name of Darwin's home. The year 2009 also marked Darwin's bicentenary; the many articles published in that twelve months—almost all dedicated to Science or Person or both—testify to the significance of the moment and the status acquired by the building, the historical figure and the scientific theories of its former owner. A curious instance is article 35, which indeed does mention both Darwin and Down House—so it was legitimately selected for the corpus—but it does not in fact deal with them as its main topic. The focus of article 35 is *The Rubaiyat* of Omar Khayyam, which happened to have been first translated by Edward FitzGerald in the same year in which *Origin* appeared. For that reason—and certainly for the great prominence that Darwin was enjoying in 2009—the English botanist was picked as a figure to mention in the review, again proving the strength of Darwin as a popularising vehicle, as if everything or everybody associated with him could somehow enjoy part of his popular success. After 2009, as was to be expected, the interest in the home decreased considerably. The few articles published in the past few years are all classifiable as Economy, although the stance has changed, and where in past years Economy meant calling for money, today the economic lexicon refers to the exploitation of the historical building as a property of the English Heritage Trust, and the short notices in *The Times* appeared in the years 2009-2013 sound more like marketing campaigns, in which trips to Downe, whether for scientific lectures or family visits, are advertised to the public.

6. 2. Lexical Analysis

Looking in detail at the lexicon of the corpus, some of the most interesting occurrences found are listed below.

1. The person of Darwin

Text	Relevant lexicon
4 (1929)	the great naturalist great man superlatively a great man both good and great gentle modest loving loved thoughtful charitable genius both good and great perhaps the most gifted family [the Darwins] ever born in England
18 (1968)	gentle, harmonious and loving family man
19 (1973)	the family man great scientist fond father
23 (1998)	world-famous occupant surprisingly conservative revolutionary [Down House] "personality" house
26 (2008)	pigeon fancier
30 (2009)	illustrious ancestor
32 (2009)	Mr Darwin, an idea, a symbol, a battle cry
33 (2009)	a revolutionary scientist
35 (2009)	gentleman-scholar

Table 5.4. The person of Darwin

The person of Darwin seems to be (re)constructed through social, historical and ideological filters.

Epithets referred to Darwin include:

- nouns: man, father, scientist, naturalist, genius;
- adjectives: gentle, modest, loving, loved, thoughtful, charitable, revolutionary;
- superlatives, etc.: great, superlatively, world-famous, surprisingly conservative, illustrious.

Evaluation is present at various levels:

- subjectivity over objectivity—individual opinions tend to be conveyed through adjectivation and choice of nouns (*great naturalist*, *revolutionary scientist*);
- popularisation over science—the number of strictly scientific articles is limited in the long (93 years) period considered; science only seems to appear at times of financial need, to underline the importance of the house, and after the financial need is gone, during the bicentenary celebrations; these, however, also inspire the use of popularising genres, like the imaginary dialogue (*Mr Darwin*), and the celebratory poetry by a related poet (*illustrious ancestor*);
- mythification or/over mystification—the building of the myth is brought forth by various factors, the main one being time: as the temporal distance from Darwin's age increases (and his theories are confirmed), so does the scientist's aura; with the arrival of the bicentenary, his position in history, science *and* the popular imaginary had become so well established that mythification even at times turns into mystification, for example in hagiographic epithets such as *gentle, harmonious and loving family man* or *gentleman-scholar*, *perhaps the most gifted family [the Darwins] ever born in England*, *Darwin was both good and great*, which distantly echo those used to describe, for instance, communist rulers of the past[14].

[14]Cf. the official English biography of Korean dictator Kim Jong Il: "Comrade Kim Jong Il […] is […] the great leader of the Workers' Party of Korea and the Korean People" (*Kim Jong Il, Brief History*: iii); "Possessed the power of keen observation, […] clear analysis and extraordinary perspicacity with regard to things and phenomena. […] he was always generous, unceremonious and warm-hearted among people" (Ibid. : 2).

2. The science of Darwin

Relevant lexicon, Text	
a well-equipped biological research station, 1	compiled, 13
Beagle diaries, 19	controversies, 41
collection of pamphlets, 13	creationist, 29
evolutionary rules, 29	discoveries, 27
evolutionary theory 19	DNA, 27, 29, 32
human nature, 4	doctrines, 4
human thought, 4	dogmatism, 27
identification and classification of species, 24	evidence, 27
interpretation of living things, 4	evolution 21, 25, 27, 28, 29, 31, 32, 38
light of reason, 4	exhibits 19, 31
modern scientists, 19	experiment*, 28, 31, 32, 38, 39
natural environment, 18	fully erect man, 26
natural selection, 25, 26, 29, 31, 32	genetics, 29, 31, 32
natural world, 29	greenhouses 1
orchid site, 14	handwriting, 13
orchid studies, 14	laboratory, 2, 5, 6, 28, 31
original manuscript, 19	manuscripts, 13, 23
personal papers, 13	mechanism, 27, 29, 32
prime impulse, 4	microscope, 4
research botanist, 25	naturalist, 4, 5, 23
scientific work, 19	notebooks, 13
species theory, 13	observations 15
specimen jars, 23	observations, 28, 29, 31, 36
studies on earthworms, 1	orchids, 1, 15
theory of natural selection, 25, 32	plant*, 14, 28, 29, 31
twelve million zoological specimens, 25	proto-hominid, 26
wild orchids, 15	random mutations, 28
Worm Stone, 19, 26	research 23, 29, 31
[Worm Stone] a device for measuring the rate at which the action of worms causes objects to sink into ground 19	revolution, 20, 27
annotations, 13	seasonality, 28
axiom, 19	superstition, 27
breeding, 28	survival, 23, 28
carnivorous, 31	theor*, 21, 25, 26, 27, 28, 29, 31, 32, 41
compiled, 13	treatise, 21

Table 5.5. The science of Darwin

Specialized lexicon (biology, botany, geology, etc.) presents with a lot of occurrences in the Down House corpus. However, many of the terms are equally specialized and non-specialized (*greenhouse*, *laboratory*), depending on the context of use. Only a qualitative, hand-made analysis can verify the specificity of the setting and assign specialized labels to the terms, which instead are returned indifferently by machine-made analyses.

Within the specialized terminology in the corpus, a prevalence of evaluative terms and expressions may be highlighted (*cradle of ideas*, *light of reason, a well-equipped biological research station*) —the result of the popular medium analysed and genres and audiences linked to it. Nonetheless, some may be deemed the result of subjective opinions by the writers, especially in times of financial need or during the 2009 celebrations, both of which somehow call for heartfelt expositions. Many of these expressions are mediated from and / or shared with philosophy and psychology (*light of reason*, *prime impulse*). This is only to be expected, given the ideological charge that Darwinian theories have had right from inception, and which continue into the present day.

In more recent texts, more specific, specialized and "correct" (in professional terms) terminology starts to appear (*specimen jars*, *theory of natural selection*, *research*, *specimen*), exploiting the increased scientific education of broadsheet readers. Not only this: concepts not existing in Darwin's time or even just a few decades ago are now present (cfr. *DNA*, *genetics*, etc). A better educated readership is definitely to be taken for granted, although the question remains of whether today's readers are actually more knowledgeable in science or they confuse their familiarity with technology— the manifestation of science best known by the general public—with scientific erudition. Another interesting question is whether the average *Times* reader is also as versed, for example, in economics or in law, which definitely need some knowledge, at least to follow the financial vicissitudes of Down House in the first half of the twentieth century.

3. The economy of Darwin ('s house)

Relevant lexicon, Text	
Assessment Committee, 6	heritage, 31, 37
claim exemption, 6	issue, 6
claiming the rates, 6	items, 7, 20
full rateable value, 6	jurisdiction, 6
interactive audio tour, 38	maintenance, 1
non-payment of rates, 5, 6	million, 25, 26, 29, 31, 32
paid tribute, 12	museum, 26, 27, 32, 33, 36, 38
plead exemption, 6	offers, 3, 20
points of law, 5	owners, 12
Quarter Sessions, 6	paid, 6, 37
rating authority, 6	possession, 12
to pay rates, 76	price, 1, 37
to split the assessment, 6	properties, 3, 6, 7, 16
visitors a year, 23	prosecutors, 6
acquire, 16	purchase*, 2, 3, 5, 37
adjourned, 5	raised, 1
admission, 33, 37, 41	rates, 6, 12
assessed, 6	rating, 6
auction, 3	realise, 24
bought, 14, 20	recreation, 41
case, 5	refurbishing, 21
cheaper, 6	reopening, 27
clients, 6	renovations, 33
collection, 26, 36	restoration, 20, 21, 22, 23
cost 1, 16	revamp, 29
courts 6	safari, 41
curate*, 31, 33	sale, 24
defendants, 6	sixpence, 19
display, 26, 34	sixpenny, 7
endowment fund, 1	spend, 37
exhibition, 26, 27, 29, 33, 41	summons, 6
fetch, 1, 24	tour*, 28, 31, 38, 39, 40
free, 8, 9, 10, 11	valuation, 6
fund, 1, 21	value, 1, 27, 37
gift, 7, 13	visit, 28, 33, 37, 38, 40
grant, 16, 21	voucher, 37
guide, 41	worth, 20, 27, 32

Table 5.6. The economy of Darwin's house

The surprisingly (or is it?) specialized, accurate and consistent economic and legal terminology employed in describing the tormented history of the property is, as indicated above, a characterising feature of the corpus. This, of course, makes one wonder about the kind of readership the newspaper has enjoyed in the 1927-2012 period considered, and how it has evolved in almost ninety years. One satisfactory answer could be that *The Times*, defined "elite" (Bell 1991: 104), "highest-grade" and "upmarket" in the late 20th century (Bell 1991: 109), at the beginning of the 20th century— from which most of the Economy articles are and when Down House underwent its most difficult period—circulated among an even more restricted number of as educated readers. Only the upper class and bourgeoisie were indeed privileged enough to have the competences, the interest and the monies required to understand, desire and purchase quality newspapers. It is also interesting to notice how the economic and legal lexicons have undergone no significant diachronic variation, except in the specific currency terms (*sixpenny*). This is in line with the conservative nature of the professional communities of economists and lawyers, which tend to set exclusive thresholds against non-specialists, one of which is indeed their specific language.

The changes occurred to Down House throughout the decades can be summarised as follows:

- from an object with real estate market value to a symbol, thus
- from assessable to invaluable, and
- from a residence to a museum.

It is therefore now a product and, as such, a process of marketing and promotion is shown in the corpus to have been initiated and become fully operative in time: *pounds* have given way to *visitors-a-year* as a unit of measure. The discourse of the texts in the corpus becomes increasingly promotional (Bhatia 2004), and increasingly hybrid, incorporating rhetorical and marketing devices (*a wonderful way to spend a summer's day*, *2 for 1 entry at English Heritage*, *interactive audio tours*, *safaris*, *vouchers*). The "great complexity of the popularizing system" Gotti (2013: 28) and of the new, globalised world is reflected in the evolution of newspaper article genres and language, and requires new, up-to-date tools of investigation to interpret them.

7. Conclusions: Darwin: Man, Myth, Money

Down House may seem like a "frivolous" factor to look at in a linguistic survey of scientific popularisation, and the Down House corpus is only a small sample collection from the wider Dar. Bri. N corpus under construction.

Yet, its analysis has returned at least a few results of historical, socio-economic and cultural—if not scientific (but the dissemination level, at least, *is* there) —interest, which are thought to represent a significant basis for an exploratory investigation into the popularisation of Darwin (ism). Preliminary conclusions that can be drawn from the analysis are as follows.

The horizon of expectation in selecting the keywords for this mini-corpus was that the string "DOWN HOUSE" could retrieve texts dealing with both Darwin's figure and Darwin's science. In fact, the balance was heavily tipped in favour of the former. Indeed, the object of the popularisation process generally appears as hybrid and ambiguous, focussing on the persona of Darwin—not the person, and the ideologies surrounding his science—not the ideas behind them. In other words, it emerges that "popularisation" is to be understood as "making popular"—with popular in its acceptation of "Liked or admired by many people, or by a particular person or group"[15]—rather than in its more neutral sense of "dissemination".

In particular, considering the three main semantic areas identified—1) the person of Darwin, 2) the science of Darwin and 3) the economy of Darwin—1) and 3) have returned texts mostly dealing with "making Darwin popular", often in the sense of "making him *successful* with the general public"; on the other hand, 2) has produced texts closer to the idea of popularisation proper, i.e. the dissemination of science among non-experts. However, the latter cannot be said to reflect the corpus's main aim in terms of either quantity or quality, since any scientific knowledge on Darwinism is either taken for granted, or simply "reminded" to the reader through vague references and quotations of key words not defined or explained.

As regards the popularising strategies adopted in the texts, these, too, quantitatively reflect the main purpose highlighted of making Darwin, rather than his science, popular. This is clear in the two main strategies identified:

[15] *Oxford English Dictionary*, s. v. POPULAR.

- mythification at the conceptual level—by way of transfer:
  - the man gets objectified and even sanctified (*For he is no longer a man. He is an idea, a symbol, a battle cry*, t. 32);
  - the object (the house) gets personified (*it wasn't just a home*, t. 28; *This is English Heritage's first 'personality' house*, *Charles Darwin's restored home hopes for 40, 000 visitor [sic] a year*, t. 23; *the place was always remembered with such vivid affection by so many*, t. 19);
- mythification at the linguistic level—through promotion:
  - promotional discourse and genres are widely employed, which include rhetorical strategies and uses like direct questions and imperatives (*2 for 1 entry at English Heritage*, *Why not take a trip to the home of Charles Darwin, Down House* [...] ? t. 37; *learn about science and evolution in an interactive audio tour narrated by Sir David Attenborough*, t. 38);
  - at the lexical level, the semantics of mythification is prominent, with specific words from semantic areas between myth, money and (mild) scientific dissemination (*Down House has been a place of **pilgrimage***, *He had wanted to create a **dynastic** seat*, *This was his **nest***, t. 23; *To us, everything at Down was perfect. That was an **axiom***, t. 19—bold added).

It ought to be remembered that the above is a purely lexical analysis, focusing on semantic aspects, and with a specific interest for the social implications and power relations conveyed by the corpus, as in CDA. Many more types of analysis could be applied to the texts, and many more features could be used to cross-check them. To mention but a couple, it would be interesting to classify and study the texts according to the sub-genres and sections of *The Times* they belong to, in order to fine tune the purposes of each. Creating frequency lists and keywords could also highlight aspects not previously considered, and so could looking at the visual aspects (photographs, drawings, layout, etc.) of the articles. In brief, where this is only a sample investigation, to obtain a thorough analysis of this or of any other corpus of texts dealing with science popularisation, an integrated, multidisciplinary approach would be necessary, welcome and desirable, to "favour the adoption of close interdisciplinary contact and integration, and rely on the methods and findings of the research traditions of different fields" (Gotti 2013: 29). Nonetheless, it is worth underlining how even this limited lexico-semantic analysis came up with indications as to the social trends and power relations underlying the discourse of Darwin and Down House, confirming the crucial role of lexicon—not "just words"—in dissemination and popularisation, and of CDA as a tool of investigation.

7. 1. Present and Future Challenges

As hinted at in its description, the Dar. Bri. N is an ample, ongoing project. It is a rich and constantly expanding corpus that still needs to be thoroughly analysed, featuring a wide range of genres, sub-genres and hybrid genres that offers multifarious opportunities for research on various aspects of Darwin (ism) and from all scholarly perspectives, including multidisciplinary ones.

In particular, in terms of quantity, the long timespan the Dar. Bri. N covers, obviously increasing by the day, means it can be broken down chronologically into smaller corpora as needed and according to various criteria: date, publication, audience, etc. Furthermore, the size of the corpus is increasing not just due to time passing, but because there is a growing number of publications and many of them are also going online, to a medium that allows for much more space and thus information. This poses a methodological challenge, in that medium and genre hybridisation represent, at least for linguistic studies, an added layer to unravel, but it is true that more powerful tools for corpus analysis are being developed accordingly—though machines still currently need significant human backup[16].

In terms of quality, the Dar. Bri. N project can be of interest to many because the initial search key is DARWIN alone[17], thus the articles included cover diverse subjects, from science to tourism to the arts: individual research paths may thus be followed, and sample investigations (like the present one) may be carried out, both as standalone and as sub-parts of larger investigations. The topicality of updating and expanding this collection is guaranteed by the continuing public interest in Darwin's life and accomplishments (as occurred in 2009 for his bicentenary and 150th year since the publication of *Origin*) and, in general, by the still unchallenged supremacy of "technology as science" perceived by the population at large, which is reflected in a steady interest for the pioneers of modern science.

---

[16] The small Down House corpus itself features 25 texts in PDF and 16 more in digital format, with clear problems for any machine analysis.

[17] Of course, meaning Charles Darwin *in primis*, and not (only) his descendants-and certainly not the Australian city, which frequently "contaminates" results.

The work on the Dar. Bri. N progresses, among challenges posed by the evolution and hybridisation of cultures, languages, media and genres—in a process not dissimilar from the evolution and hybridisation that all the species underwent in their development and that Darwin, strenuously at first but successfully in the long term, managed to make acceptable to his fellow scientists and popular among the general public.

CHAPTER SIX

THE POPULARISATION OF SCIENTIFIC DISCOURSE FOR THE ACADEMIA FROM A DIACHRONIC PERSPECTIVE: THE CASE OF NOBEL LECTURES

STEFANIA M. MACI

1. Introduction

Amongst the definitions given by the OED for the term "popularisation", the one used in specialized contexts and yet still attributed with a negative connotation, refers to "the presentation of a technical or specialized subject in a generally intelligible or appealing form". Indeed, popularisation is widely perceived as the practice of disseminating scientific or technical findings outside the communities within which such findings are produced and which, therefore, "own" such knowledge (Giannoni 2008: 212). In other words, the core meaning of popularisation seems to lie in the distorted belief that only experts have access to genuine scientific knowledge, whereas popularised knowledge is merely a simplification of specialized knowledge for non-specialists (Myers 2003: 265; Hilgartner 1990: 519). This implicitly supports the existence of an unequal power relationship between experts and non-experts, since "real" knowledge is assigned to experts (Myers 2003: 266). From this distorted and naïve viewpoint, the existence of specialized knowledge (within the scientific community) and lay knowledge (outside the scientific community) seems to be reproduced in two types of discourse, one within and one outside scientific institutions. In this context, popularisation is realized as a translation from one type of discourse to another. Viewed from this perspective, popularisation is seen simply as a re-contextualization and reformulation of a (scientific) source text in order to provide varied audiences with greater comprehension and accessibility (de Oliveira/Pagano2006: 626; Ciapuscio 2003: 210). These audiences can thus elaborate lay versions of scientific knowledge and integrate them with their own existing knowledge. However, if popularisation were simply a matter of recontextualisation, not only would a distinction between scientific and popularised knowledge be required, but also a clear-cut identification of the borders existing between "genuine" scientific knowledge and "popularised" representations. This in turn would entail a definition of expert and lay audiences. Yet as Hilgartner (1990: 524-525, 529) claims, such distinctions are anything but clear. On the contrary, the in-depth analysis of popularisation recently carried out by the applied linguistics field, [1] has revealed that popularisation cannot merely be seen as a distorted simplification of scientific knowledge for non-specialists: popularisation may also be seen as a matter of degree which operates along a continuum, going from researchers to an educated public and with practitioners positioned somewhere in the middle (Giannoni 2008; Myers 2003; Hiltegartner 1990). The contexts within which this continuum operates range "from laboratory 'shop talk', to technical seminars, to scientific papers in journals, to literature reviews, grant proposals, textbooks, policy documents, and mass media accounts" (Hiltgarner 1990: 524). In addition, because of the action of the different agents, voices and genres, involving different actors, institutions, and forms of authority, the diffusion of scientific texts among members of the scientific community stepping outside their very limited specialist area can also be regarded as popularisation (Myers 2003: 267).

It is clear, therefore, that the term "popularisation" has ambivalent meanings. This, therefore, calls for some sort of disambiguation. On the one hand, we have texts which are used to present scientific knowledge in a readable and meaningful way, addressed to non-specialist readers; on the

---

[1] The language of popularisation has attracted recent interest in the field of linguistics, focussing on defining the scope and boundaries of the discourse of *popularisation* (Leake 2012; Giannoni 2008; de Oliveira/Pagano 2006; Calsamiglia/van Dijk 2004; Paul 2004; Calsamiglia 2003; Gotti 2003; Myers 2003; Cooter/Pumfrey 1994; Thoiron/Béjoint 1991; Hilgartner 1990; Jacobi 1990). Other investigations have concentrated on the role played by the media in popularisation when bridging the gap between scientific knowledge and lay experience (Dossena 2008; Ciapuscio 2003; Moirand 2003; Beacco *et al.* 2002; Barton 1998), which can be expressed even through metaphorisation (Pramling/Säljö 2007; Knudsen 2003; Skorcynska 2001; Leane 2001). Research has also focussed on the language of popularisation for pedagogical purposes (Parkinson/Adendorff 2004, 2005; Sharma 1972).

other, there are those types of texts used by expert members of the scientific community to disseminate scientific knowledge among less expert members and/or across specialisations. We can therefore say that the discourse of popularised science includes the former range of texts, whereas the discourse of scientific popularisation is the umbrella term comprising the latter class of texts. Whether a text belongs to the discourse of either popularised science or scientific popularisation depends on its discourse target and purpose (Gotti 2003). Table6.1., below, summarises these concepts:

	Discourse of popularised science	**Discourse of scientific popularisation**
Purpose	Knowledge construction	Knowledge dissemination
Target	Lay people, wider public	Members of the scientific community
Text type	Mass media accounts; web documents (blogs; forums); government policy documents	Laboratory talk; grant proposals; project delivery;textbooks; mass media accounts; web documents (blogs; forums); policy documents

Table 6.1. Discourses of popularisation.

Within applied linguistics there is, however, a lack of attention paid to research when specifically centred around the way in which discourse is "popularised" by expert members of the scientific community for the sake of professionals whose expertise may not be within the same field as that of the text authors. To the best of my knowledge, the only studies carried out in this sense are respectively the one by Giannoni (2008) and by Paul (2004), which analyse this topic in a synchronic way. It is therefore the aim of this research to disclose the extent to which scientific knowledge is rendered as scientific popularisation by members of the medical academic community for an academic audience stepping outside the same professional category and across time. More precisely, I will investigate in what ways this occurs in such prestigious academic events as the presentation of Nobel Prize Lectures on Physiology or Medicine.[2] By

[2] The Nobel Prize in Physiology or Medicine, as described in Alfred Nobel's will, is dedicated to "the person who shall have made the most important discovery within the domain of physiology or medicine" (http: //www. nobelprizemedicine. org/?pageid=2266 [18/05/2013]). Nominations are invited from scientists from all over the world and are evaluated by the Medical Nobel Committee, which

comparing 20 Nobel Prize Winner Lectures (NLs), with the BASE corpus of academic lectures on Life and Medical Sciences,[3] I will use a Corpus Linguistics approach in order to detect how popularisation has been realized across time, despite any possible professional-related conditioning effects. The relative results will be discussed in the following paragraphs.

2. Methodological Approach

The present study analyses twenty NLs given by winners of the Nobel Prize for Medicine or Physiology, as listed in Table 6.2. below. The twenty NLs were chosen at random in the period ranging from 1901 (the year when the Nobel Prize was first established) to 2010 (the year when the most recent Nobel Prize was awarded in respect to the date my research began). During these years one or two NLs per decade were selected. [4]

---

identifies the leading candidates. The Nobel Assembly, which consists of 50 professors from the Karolinska Institutet, makes the final decision by vote (http: //www. nobelprizemedicine. org/?pageid=330 [18/05/2013]) and confers the award, usually on the first Monday in October. (http: //www. nobelprizemedicine. org/?pageid=2266 [18/05/2013]). According to the Nobel Foundation statutes, the Nobel Laureates are required "to give a lecture on a subject connected with the work for which the prize has been awarded" (http: //www. nobelprize. org/nobel prizes/medicine/videolectures. html [18/05/2013]). Laureates in Physiology or Medicine give their Nobel Lectures at the Karolinska Institutet, usually on the 7th or 8th of December each year (http://www. nobelprizemedicine. org/?pageid=2673 [18/05/2013]). All NLs are theoretically open to the general public. It is usually rather difficult to manage to attend the lectures on medicine, since the main auditorium at the Karolinska Institutet is quite small (email communication by Jonna Petterson Informatör/ Public Relations Officer Nobelstiftelsen, Oct. 8, 2012). For this reason, they are also broadcast by video link to two other adjacent lecture halls. According to Tatiana Goriatcheva, Administrator of the Nobel Committee for Physiology or Medicine (email communication, Oct 18, 2012) the audience mainly consists of professors, scientists and research students from the Karolinska Institutet as well as Stockholm University, and the Nobel Laureates know in advance what type of audience they will be addressing.

[3] Available at http: //www2. warwick. ac. uk/fac/soc/al/research/collect/base/ [18/05/2013] .

[4] The decision to select one or two NLs depends on the fact that, during the First and Second World Wars, Nobel Prizes were not assigned. Information available at www. nobelprize. org.

Decade	Year	Nobel Winner	Nobel Lectures (NLs)	
1901-1910	1901	Emil von Behring	*NL1*	*Serum Therapy in Therapeutics and Medical Science*
	1905	Robert Koch	*NL2*	*The Current State of the Struggle against Tuberculosis*
1911-1920	1914	Robert Bárány	*NL3*	*Some New Methods for Functional Testing of the Vestibular Apparatus and the Cerebellum*
	1920	Schack A. S. Krogh	*NL4*	*A Contribution to the Physiology of the Capillaries*
1921-1930	1924	Willem Einthoven	*NL5*	*The String Galvanometer and the Measurement of the Action Currents of the Heart*
	1930	Karl Landsteiner	*NL6*	*On Individual Differences in Human Blood*
1931-1940	1935	Hans Spemann	*NL7*	*The Organizer-Effect in Embryonic Development*
	1939	Gerhard Domagk	*NL8*	*Further Progress in Chemotherapy of Bacterial Infections*
1941-1950	1946	Hermann J. Muller	*NL9*	*The Production of Mutations*
1951-1960	1951	Max Theiler	*NL10*	*The Development of Vaccines against Yellow Fever*
	1957	Daniel Bovet	*NL11*	*The Relationships between Isosterism and Competitive Phenomena in the Field of Drug Therapy of the Autonomic Nervous System and that of the Neuromuscular Transmission*
1961-1970	1964	Konrad Bloch	*NL12*	*The Biological Synthesis of Cholesterol*
	1964	Feodor Lynen	*NL13*	*The Pathway from "Activated Acetic Acid" to the Terpenes and Fatty Acids*
1971-1980	1972	Gerald M. Edelman	*NL14*	*Antibody Structure and Molecular Immunology*

	1972	Rodney R. Porter	*NL15*	*Structural Studies of Immunoglobuline*
1981-1990	1986	Stanley Cohen	*NL16*	*Epidermal Growth Factor*
	1986	Rita Levi-Montalcini	*NL17*	*The Nerve Growth Factor: Thirty-Five Years Later*
1991-2000	1996	Peter C. Doherty	*NL18*	*Cell Mediated Immunity in Virus Infections*
2001-2010	2009	Elizabeth H. Blackburn	*NL19*	*Telomeres and Telomerase: The Means to the End*
	2009	Carol W. Greider	*NL20*	*Telomerase Discovery: The Excitement of Putting Together Pieces of the Puzzle*

Table 6.2. List of NLs

All the NLs examined were downloaded from http://www. nobelprize. org/nobelprizes/medicine/. Copyright permission to use and reproduce the NLs was granted by the Nobel Prize Foundation. [5] The documents were saved in pdf format and later transformed into text files, forming a corpus of 125, 876 words, for use in the concordance of software programs. The quantitative (Biber et al. 2007; Dörnyei 2007) and qualitative (Coffey/Atkinson 1996; Miles/Huberman 1994) analyses carried out in my research are based either on automatic or manual searches, or a combination of both. The reference corpus is the BASE (British Academic Spoken English) corpus of the University of Warwick and Reading and was collected by H. Nesi and P. Thompson It is available at http: //www2. warwick.ac.uk/fac/soc/al/research/collect/base/ and comprises 160 academic lectures, from which the 40 related to "Life and Medical Sciences" section were selected (cf. http://www2.warwick.ac.uk/fac/soc/al/research/collect/ base/lecturetranscripts/ls). The text files were then read and checked manually against any written element related to the BASE tagging which could interfere with the corpus analysis. The resulting edited corpus comprises 254, 267 words. For the computer-based counts, Wmatrix3[6] (Rayson 2009) search options have been used. The results, based on log-likelihood statistics ($p<0.01$) and presented in normalised figures (per 1, 000 words), were then accompanied by manual correction to rule out any non-relevant cases. For a classification of key speech parts in the

[5] Personal email received on April 12, 2012.

[6] Available at http://ucrel. lancs. ac. uk/wmatrix3. html.

corpus, I scanned the target texts twice: the first time to locate any occurrence of such features; the second time to verify their actual status on contextual and pragmatic grounds.

3. Lectures and NLs: Preliminary Findings

Lectures are an instructional genre in higher education (Lee 2009; Kiewra 2002; Miller 2002; Flowerdew 1994) whose pragmatic functions have recently been identified by Deroy and Taverniers (2011) in a study taking into account the overall communicative purposes of the lecture genre. These are shown in Table 6.3. below:

Functions	Subfunctions
Informing	Describing Recounting Reporting Interpreting Demonstrating
Elaborating	Exemplifying Reformulating
Evaluating	Indicating attitude Indicating degree of commitment
Organizing discourse	Orientating Structuring Relating
Interacting	Regulating interaction Involving the audience Establishing a relationship with the audience
Managing the class	Managing organizational matters Managing delivery Managing the audience

Table 6.3. The communicative purposes of lectures (from Deroey and Taverniers 2011: 5)

As Deroey and Taverniers (2011) claim, the *informing* function is realised by the fact that lectures are mainly a means of disseminating subject information to students. The *elaborating* function is realised when the understanding of a lecture takes place, thanks to the help of the lecturer. In this sense, lectures offer the lecturers' own representation of the subject world (the *evaluating* function). In order to do so, lectures are pre-planned by the lecturers, who guide their listeners through the complex and dense informative message decoded by the listeners in real time (the *discourse organization* function). This implies that a sort of interaction is established between the lecturers and their audience (thus complying with the *interacting* function). Finally, lectures are events "managed in regard of its organization, delivery and audience" (*managing the class*) (Deroey and Taverniers 2011: 16).[7]

All the NLs forming my corpus apparently have most of the same communicative purposes because their main aim is that of:

(i) disseminating knowledge among the scientific audience not belonging to the same professional category as that of the Nobel lecturer (*informing*);
(ii) helping the audience to understand the NL (*elaborating*);
(iii) offering a representation of the world studied by the Nobel Prize lecturer (*evaluating*);
(iv) guiding the audience towards decoding the scientific message (*discourse organization*);
(v) establishing interaction between the lecturers and the audience (*interacting*).

The only communicative purpose which is absent in NLs is the *managing the class* function-clearly, the audience at the Karolinska Institutet does not need to be managed.

Among the above-mentioned functions there are, however, certain sub-functions which are missing, as can be seen in Table 6.4. below, where only the relevant sub-functions are indicated:

Functions	**Sub-functions**
Informing	Describing Recounting Reporting
Elaborating	Exemplifying
Evaluating	Indicating degree of commitment
Organizing discourse	Orientating Structuring Relating
Interacting	Establishing a relationship with the audience

Table 6.4. The communicative purposes of NLs (adapted from Deroey and Taverniers 2011: 5)

[7]It is not my purpose here to explain in detail the functions and sub-functions of lectures. For an in-depth account of these, see Deroey and Taverniers (2011).

The fact that the *interpreting* and *demonstrating* sub-functions in the *informing* function, the *reformulating* sub-function in the *elaborating* function, the *indicating attitude* in the *evaluating* function and the *regulating interaction* and *involving the audience* in the *interacting* communicative purpose are missing is quite understandable: scientists do not need them because the findings they are describing are the reason why they have been awarded the Nobel Prize-their discovery is now commonly-shared knowledge in the medical field.

The main issue in NLs is that they are "lectures", that is, instructional tools targeting a professional audience whose professionalism steps outside the same professional sphere as that of the Nobel Prize winner. In the introduction to his 1914 NL, Robert Bárány claims:

> Naturally I can only present a fraction of these results and I shall have to content myself with telling you briefly the history of those discoveries, the understanding of which only requires a minimum of specialist knowledge. I have already given, on repeated occasions, a historical account of my discoveries. I must, therefore, ask the indulgence of those specialist colleagues, the Prize judges and the members of the Royal Caroline Institute who have found my work worthy of the highest scientific distinction if I say nothing which is new or interesting to them. This, if I have understood it right, is not the purpose of the Nobel Lecture either, for it should serve far more to give the lay public a glimpse into the world of thought and work of the research scientist, just as the distinguished founder of this prize was a layman himself. To raise the general cultural level of mankind by promoting scientific work lay very close to the heart of Alfred Nobel. I shall ask this eminent gathering to regard my lecture from this standpoint. (NL3)

NLs have therefore always been recognised as lectures belonging to the discourse of scientific popularisation.

In order to see whether any syntactical differences exist between NLs and academic lectures, I used WMatrix online software to compare the two corpora on the basis of their parts of speech (POS). [8] The first three key POSs have evidenced two clear differences between them, as revealed by

[8] WMatrix is a type of software elaborated by the UCREL (University Centre for Computer Corpus Research on Language, Lancaster) which automatically tags words at their lexical, grammatical and semantic level. Computation is based on log-likelihood statistics ($p<0.01$), and shows key items at the top. Part of Speech (POS) tagging is generated by the CLAWS (the Constituent Likelihood Automatic Word-tagging System), tagger, which has been continuously developed since the early 1980s. The latest version of the tagger, CLAWS4, was used to POS tag c. 100 million words of the British National Corpus (BNC).

Table 6. 5. below, where we find the top key POS for both the BASE lectures (above) and the NLs (below). In the first column the numbers "1", "2" and "3" indicate the ranking position and the expressions "List1", "List2", etc., are links redirecting to the list of the terms found; they are followed by the *concordance* of the part of speech under consideration; the column entitled "Item" presents some acronyms referring to the class of the POS under consideration. The column named "O1" refers to the observed frequency in my selected corpus of BASE lectures (above) and NLs (below). "O2" refers to the observed frequency in the reference corpus, which will be the NL corpus when the BASE corpus is under investigation; vice versa, it will be the BASE corpus when the NLs are taken into consideration. The values of %1 and %2 show relative frequencies in the texts; the symbol "+" indicates over-use in O1 relative to O2, while the symbol "-", if present, indicates under-use in O1 relative to O2. The two tables are sorted on log-likelihood (LL) value to show key items at the top.

As to the acronyms shown in the column "Item", we are not going to look at Table 6. 5. in detail because for the purposes of this analysis only rows 1-3 are of interest. As to the Lecture key POS, the acronym "PPY" refers to the person pronoun "you"; "FU" to any unclassified word (which in our case corresponds to the hesitation "er" typical of oral discourse);"VBZ" to the verb "to be" in the third person singular. As to the NL key POS, "NP1" refers to singular proper nouns (which in our case corresponds to the names of the Nobel Prize winners); "PPIS1" to the person pronoun "I"; "VVD" to lexical verbs (expressed in the past form).

In other words, simply by looking at rows 2 and 3 of both the BASE corpus and the NL corpus, we can say that the main difference lies in the fact that the BASE corpus seems to be more instructional or assertive than the NL corpus given the presence of the verb "to be" as a keyword in the Lecture corpus. Indeed, not only is the BASE corpus apparently characterized by the exploitation of the phatic function indicating the attempt to check whether communication is successfully achieved-hence the keyness of the pronoun "you"; but also by definitions introduced by the existential verb "to be".

On the other hand, the NL corpus seems to evidence a completely different type of narrative, where the researcher is the main protagonist of the research itself. The researcher acquires the subject-position, which clearly indicates that the author is the actor of the research. The verb tends to be in the past form, as in the example above where the authors *observed* the phenomenon, as indicated in Figure 6.2. below.

Lecture key POS										
				Item	**O1**	**%1**	**O2**	**%2**		**LL**
1	List1	List2	Concordance	PPY	5580	2.24	3	0.01	+	985.77
2	List1	List2	Concordance	FU	5674	2.28	11	0.05	+	933.71
3	List1	List2	Concordance	VBZ	7736	3.11	194	0.81	+	539.48
NL key POS										
				Item	**O1**	**%1**	**O2**	**%2**		**LL**
1	List1	List2	Concordance	NP1	810	4.76	1368	0.55	+	1757.68
2	List1	List2	Concordance	PPIS1	122	0.72	2	0.00	+	650.23
3	List1	List2	Concordance	VVD	412	2.42	1435	0.58	+	493.54

Table 6. 5. Top three key POS in lectures and NLs

The narrative pattern in the two corpora is therefore different because of the different pragmatic functions exerted by them: instructional in the BASE corpus, storytelling in the NL corpus. This story has a beginning, a complicating action, a resolution and evaluation (in much the same way as indicated by Labov's schemata (1972, 1997, 2001, 2002) of fully-formed oral narratives)[9], and such pattern has been maintained across time, as indicated in Table 6.6 below, where the most relevant examples are reproduced.

[9] According to Labov, oral narratives are structured in the following parts: *Abstract*, *Orientation*, *Complicating action*, *Resolution*, *Coda* and *Evaluation*, with the latter pervading all the other components.

5580 occurrences.			Ex
so on the handout	you	've got this and i do n't i 'm not	1
nd and start talking about it but	you	have got this and hopefully you 'l	2
t you have got this and hopefully	you	'll see this develops come along s	3
other side of your sheet is trash	you	can pretty well ignore it 's about	4
hanged it slightly and i 'll give	you	another sheet er just now but but	5
eet er just now but but basically	you	've got from this week after the l	6
week after the lecture i 'll give	you	a a a list and some information to	7
e information to pick a topic and	you	each pick one topic you 'll do you	8
topic and you each pick one topic	you	'll do your poster and your essay	9
and your essay on that one topic	you	'll then prepare it during the tim	10
and then in weeks eight and nine	you	'll present the posters before we	11
ent the posters before we 'll ask	you	questions about and et cetera and	12
posters in week seven er er yeah	you	'll need to give us the er the pos	13
we can get it er in a state that	you	can present it on the on the next	14
t to do is to start off by giving	you	some of the general introduction t	15
plant breeding itself okay now if	you	do have questions then please do f	16
to please ask whatever questions	you	want to okay but otherwise just as	17
g okay as i said i 'm not sure if	you	were all here when i said what i '	18
'll try and do is to try and give	you	handouts of all the overheads so e	19
uts of all the overheads so er so	you	should n't have to scribble them d	20
cribble them down too much but er	you	may want to add more information i	21
the other important thing to tell	you	despite what it says on your timet	22
to s- , start by doing and i hope	you	do n't regard this as too trivial	23
far as humans are concerned what	you	have to remember is that we are co	24
s we go on to talk about cropping	you	do realize that central role and a	25
ing about quite narrow things but	you	need to bear in mind the broader c	26
oit them okay so having said that	you	'll not be surprised if when we as	27
hat was growing particularly well	you	'd not be surprised if his neighbo	28
by this picking of the best then	you	'd start to get qu- , fairly quick	29
rowing around them in other words	you	'd start to get domesticated forms	30

7736 occurrences.			
ion this week and then there	's		the main types of breeding sc
the other side of your sheet	is		trash you can pretty well ign
ou can pretty well ignore it	's		about the er the posters and
r and then week ten if there	's		anything to pick up we can do
me back to that okay so that	's		the main outline of things wh
of things what i want to do	is		to start off by giving you so
i said what i 'll try and do	is		to try and give you handouts
etable next week the lecture	is		also here because they have n
t regard this as too trivial	is		just to put plants in context
ed what you have to remember	is		that we are completely depend
ing to exploit them and food	is		not the only product we get f
it about food crops but that	's		not because they 're the only
se they 're the only ones it	's		just it 's it 's slightly eas
the only ones it 's just it	's		it 's slightly easier to talk
nly ones it 's just it 's it	's		slightly easier to talk about
g and for our survival so it	's		important when we go on to ta
ing been going on the answer	is		as long as we 've been practi
cognized this so i mean that	is		is the case and i 'm not er i
nized this so i mean that is	is	the case	and i 'm not er in any way be
tling their contribution it	's		very very big and they did it
riving force and a major one	is		increasing productivity certa
most of the crop plants that	is		still a major consideration t
o produce more per unit area	is		really just as it 's always b
if you think about it there	's		really no industry no i- , sy
efficiency and productivity	is		not sensible you 'd be better
ound profitability et cetera	is		still an important considerat
pact that we 're having this	is		something else that we have t
ich are suited to that there	's		no point thinking you can jus
r the increase in production	is		still there and is still a ma
roduction is still there and	is		still a major driving force t

Figure 6.1. Concordance lists of *PPY*and *VBZ* in the BASE Corpus.

122 occurrences.

from unlikely places . In this paper	I	describe the process of scientific dis
DEFINING THE EDGES OF THE PUZZLE When	I	joined Liz s lab in May of 1984 I set
hen I joined Liz s lab in May of 1984	I	set out to look for this unknown enzym
d but not at the other (figure 4) .	I	incubated this linear fragment of DNA
end than the end lacking a telomere .	I	incubated the linear DNA substrate in
sors . After incubation for an hour ,	I	purified the linear fragment from the
e ends . Following its purification ,	I	cut the DNA fragment to generate two u
nts ? After our initial attempts that	I	ve described above , we sat and puzzle
red in length by just a single base .	I	worked from May through December tryin
stead of a long linear DNA fragment ,	I	tested a synthetic 18 residue oligonuc
his synthetic oligonucleotide , which	I	decided to use instead of the DNA rest
was seen on December 25 , 1984 . So ,	I	set out to examine the elongation of t
on of the telomeric oligonucleotide .	I	made cell extracts from Tetrahymena an
generated by the radioactive label ,	I	exposed the gel to x-ray film for thre
l to x-ray film for three days . When	I	went to the lab to develop the x-ray f
o the lab to develop the x-ray film ,	I	was thrilled to see a repeating patter
REALLY FIT , OR ARE WE FORCING THEM ?	I	talked to Liz the next day and showed
DNA . These were exciting times. once	I	could repeatedly see the primer elonga
eas about how it might be generated .	I	would come in to the lab every day , e
GGG repeats was coming from . Liz and	I	talked through several different model
gh several different models . The one	I	wanted most to test was that there mig
quences added to the chromosome end .	I	set out to do an experiment to pre-tre
Berkeley to give a seminar . Liz and	I	met with Tom in the morning and descri
oteins present in the crude extract .	I	established a multistep purification p
column chromatography (figure 7a) .	I	would separate the extract into fracti
he fraction containing telomerase) .	I	would then take the active fractions a
another , different separation step .	I	used size exclusion , ion exchange , d
ely purify telomerase (figure 7B) .	I	then examined the active fraction to l
present when telomerase was active .	I	purified the RNAs from active fraction
n a sequence for a number of months ,	I	decided to take a more direct approach

412 occurrences.

h patience and determination ,	came together	to tell us something new about
HE PUZZLE : TELOMERE SEQUENCES	DEFINED	Telomeres posed a puzzle for bi
RE SEQUENCES DEFINED Telomeres	posed	a puzzle for biologists for a m
Muller and Barbara McClintock	showed	that natural chromosome ends ha
s have special properties that	afforded	their protection ; they coined
forded their protection ; they	coined	the word telomere to describe t
. The puzzle of how these ends	functioned	remained unsolved for many year
ains 40 , 000 telomeres . They	found	that the chromosome end was mad
this simple repeated sequence	turned out	to be the key to understanding
ss a wide variety of organisms	suggested	that telomere function might al
otecting them from destruction	arose	early in evolution , such that
to know how the simple repeats	functioned	as telomeres to confer protecti
d Gall , 1978) . How was this	established	? When a population of telomere
d using restriction enzymes it	generated	so-called fuzzy bands when the
the telomeres in trypanosomes	grew	longer as the cells were grown
rowth , the telomere fragments	grew	progressively (Larson et al. ,
ation . In 1974 , James Watson	suggested	that a linear chromosome should
osed by Alexi Olovinkov , Bwho	suggested	that chromosome ends might shor
Liz Blackburn and Jack Szostak	met	for the first time at a confere
interested in DNA ends . They	knew	about the curious structure of
nd their elongation , and they	saw	a way to use these puzzle piece
a long-shot experiment : they	decided	to test whether Tetrahymena tel
riment they could do , so they	forged	ahead . Blackburn and Szostak t
ahead . Blackburn and Szostak	took	a circular yeast plasmid and cu
nzyme to make it linear . Jack	knew	from his earlier experiments in
o yeast cells . However , they	wondered	whether the addition of Tetrahy
n transformed into (that is ,	added	to) yeast cells . Liz purified
, added to) yeast cells . Liz	purified	the Tetrahymena telomeres and J
Tetrahymena telomeres and Jack	ligated	them onto the linearized yeast
nto yeast , it was stable : it	replicated	and was maintained as a linear

Figure 6.2. Concordance lists of *PPIS1* and *VVD* in the NL Corpus.

Storytelling pattern in NLs		
	NL1 (1901)	NL19 (2009)
Beginning	Twenty years ago, tuberculosis, even in its most dangerous form, consumption, was still not considered infectious.	Tracing the beginnings of the interwoven stories of science can be arbitrary, as beginnings are so often lost in the mists of time. For me, arguably the story of telomeres and telomerase began thousands of years ago [...] .
Compli-cating action	But taken as a whole, all these efforts were quite dissimilar in character, though they all pursued the same aim, but chose quite different approaches to it.	I tried to identify the proteins on *Tetrahymena* telomeres, but did not succeed in this. It was others' work, initially using yeast molecular genetic approaches, that unlocked the door to telomeric proteins
Resolution	I would just like to observe in addition to this that the re-examination of our investigations, [...] has led to a confirmation of my view, and that the harmlessness for humans of the bacilli of "pearl disease" is directly proved, in addition, by inoculating humans with material from it [...] .	We discovered that short fragments of DNA, synthesized chemically as DNA oligonucleotides and herefore available in high concentrations, would get telomeric GGGGTT repeats added to their 3' ends then incubated with Tetrahymena extracts.
Evaluation	If these tuberculous people, whose numbers, as I have said, are truly considerable, are abandoned to their fate, then a great gap in the battle against tuberculosis would result from this.	Certainly, these findings and implications are taking the field of telomere and telomerase biology into realms far from the single-celled pond microorganisms in which I began this work.

Table 6.6. Storytelling pattern in NLs

Interestingly, starting from NL4 (1920), these narratives are accompanied by figures which are extremely technical and clearly aimed at a specialized audience, as we can see in the Figures 6. 3. a-b below, and which are not present in the research article where the discovery for which the two scholars won the Nobel Prize is reported:[10]

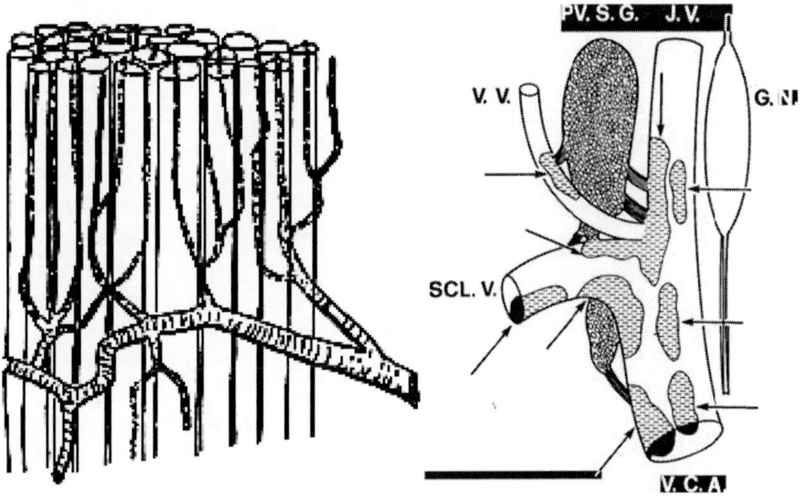

Figures 6. 3. a-b. Use of specialized illustrations in NL4 (1920, left) and NL17 (1986, right).

We can consider them as those elements which, to put it into Giunchi's words (2002: 288), "stress the local angle", i.e .the "professional" angle, in an attempt to relate discourse in popular medical texts to the real world of the reader, by explaining and in this case, describing, the concepts and principles on which the research is based. This "local angle", however, seems to be increasingly present in the NLs presented in the period from the 1930s to the 1970s. Indeed, as from 1930, there have been some formal modifications affecting NL scientific discourse: firstly, quotations within the NL began to be consistently cited following the APA citation style (1930); secondly, a reference list was provided at the end of the NL (1939); thirdly, the NL started to be divided into sections according to the IMRD (*Introduction-Methods-Results-Discussion*) sections typical of scientific academic written discourse (1957); and finally, the title of the NL began to be written *before* the Nobel Prize winner's name, as though to indicate that the research findings are more relevant than the researcher himself or herself (1976).

[10] See Krogh (1922) and Cohen, Levi-Montalcini and Hamburger (1954). Krogh published the results of his research in a series of lectures after he won the Nobel prize.

In general, readers of scientific texts exploit their knowledge of lexical relations in order to observe any implicit cohesion within the text, whereas "readers of popularised texts must see the explicitly marked cohesive relations to infer lexical relations" (Myers 1991: 1). This different exploitation of knowledge makes scientific texts harder for the non-specialist to follow. Yet, the knowledge required by NLs seems to slowly veer toward higher and higher specialization, where the gap between specialized and common usage registers does not need to be bridged. In other words, it would seem that popularised accounts of NL research tend to be more and more in line with the formal specialization technicalities typical of scientific journal articles written for specialist audiences, thus appearing to erase any existing different uses of the audience's knowledge of the two kinds of texts.

If this were so, there would be a change from explanation to a representation of how the world works. Indeed, the main purpose of specialized scientific discourse is to represent the world as it actually is, epistemically indicating the degree of commitment a speaker or writer feels towards the truth expressed by the proposition. As a consequence, this also would imply a change in the way in which *evaluation* is rendered, that is, the way of expressing the "writer's attitude or stance towards viewpoints, or feeling about the entities or propositions that he or she is talking about" (Hunston and Thompson 2003: 5). The textualized expressions of the authors' stance and opinions, within a discourse, which is thus rendered as more persuasive and effective, are normally expressed by means of hedges and boosters (Biber/Finegan 1989; Hunston and Thompson 2000; Hyland 2001). Indeed, they transform assertions into endorsed knowledge with the readers' and/or audience's acceptance, the latter being possible only with rhetorical persuasion. In other words, hedges and boosters are expressions of doubt and certainty used by authors to modify the claims they make, by toning down what is tentative and potentially hazardous and by emphasising what is thought to be exact (Hyland 2000b).

Considering the trend of NLs, in the following paragraphs I will consider whether this development from popularisation to specialisation can be detected in hedge and booster markers through which Nobel Prize winners balance caution and confidence in their statements.

4. NLs across Time: Results and Discussion

Drawing from Hyland (2000a: 188-189), I have summarised the types of boosting and hedging devices found in my NL corpus. As can be seen in Tables 6.7 and 6.8, they are usually achieved through the use of modal lexical and auxiliary verbs, as well as adjectives and adverbs. The tables indicate the frequency and Standardised Type/Token.

The ratios (STTR) [11] of hedges and boosters, which are grouped in NL sub-corpora classified during the years from 1940 to 1979, years when the most evident changes towards specialization occurred, are shown for three different periods: 1901-1939 (42, 218 words); 1940-1979 (37, 198 words); 1980-2009 (42, 205 words): [12]

	1901-1939 distribution		**1940-1979 distribution**		**1980-2009 distribution**	
	Frequency	*STTR*	*Frequency*	*STTR*	*Frequency*	*STTR*
Hedges	819	*220. 17*	314	*84. 91*	187	*44. 30*
Boosters	797	*188. 78*	342	*91. 94*	256	*60. 65*

Table 6.7. Hedges and boosters across time

As indicated in Table 6.7, NLs seem to diminish the exploitation of both hedging and boosting devices across time (see Table 6.8. below). Indeed, the 1901-1939 sub-corpus shows the highest frequency of hedges and the 1980-2009 sub-corpus the lowest, which is also confirmed by their STTR. [13] In the 1901-1939 sub-corpus, the hedging devices used much more frequently than in the 1980-2009 sub-corpus are the modals *could*,

[11] In Corpus Linguistics, *tokens* are the running words of the corpus, i.e., the number of words contained in the corpus; *types* refer to each different kind of word in a corpus. For instance, the sentence *I want what I like* is composed of five tokens, or running words, but has only four types (*I*, *want*, *what*, *like*). The TTR is the ratio between types and tokens. Since the resulting percentage may vary according to the length of the texts forming the corpus, the TTR is normally *standardised*; in other words, in order to be sure that the TTR represents fair results and percentages, the ratio is calculated for the first 10, 000 running words, then calculated afresh for the next 10, 000, and so on to the end of the text or corpus (see Hunston 2002: 17).

[12] Figures within brackets refer to the items within brackets in the first column.

[13] It is worth noting that the three sub-corpora have more or less the same number of words.

should and *would.* Adverbs such as *normally*, *often*, *perhaps*, *quite*, and *relatively* appear, together with the adjective *possible*, whereas the 1980-2009 hedge with a higher occurrence than those belonging to the 1901-1939 sub-corpus is *likely.* At the same time, boosters seem to be less frequently used across the three corpora. Once again, the 1901-1939 sub-corpus exploits boosters more than the 1980-2009 sub-corpus does.

HEDGES	**1901-1939 distribution**		**1940-1979 distribution**		**1980-2009 distribution**	
	Frequency	*STTR*	Frequency	*STTR*	Frequency	*STTR*
could	48	*11. 36*	29	*7. 79*	25	*5. 92*
likely	0	*0*	6	*1. 61*	10	*2. 36*
normally	8	*1. 89*	5	*1. 34*	2	*0. 47*
often	22	*5. 21*	3	*0. 80*	2	*0. 47*
perhaps	16	*3. 78*	6	*1. 61*	8	*1. 89*
possible	52	*12. 31*	9	*2. 41*	14	*3. 31*
relatively	10	*2. 36*	8	*2. 15*	1	*0. 23*
should	48	*11. 36*	14	*3. 76*	5	*1. 84*
would	65	*15. 39*	19	*5. 10*	13	*3. 08*
BOOSTERS	**1901-1939 distribution**		**1940-1979 distribution**		**1980-2009 distribution**	
	Frequency	*STTR*	Frequency	*STTR*	Frequency	*STTR*
actually	6	*1. 42*	1	*0. 26*	1	*0. 23*
certain	48	*11. 36*	6	*1. 61*	2	*0. 47*
clear	11	*2. 60*	5	*1. 34*	6	*1. 42*
conclude	6	*1. 42*	1	*0. 26*	0	*0*
decided	4	*0. 94*	2	*0. 53*	0	*0*
definite	7	*1. 65*	0	*0*	1	*0. 23*
evidence	4	*0. 94*	11	*2. 95*	26	*6. 16*
the fact that	29	*6. 86*	12	*3. 22*	2	*0. 47*
find / found	10 / 52	*2. 36 / 12. 31*	3 / 39	*0. 80 / 10. 48*	0 / 12	*0 / 2. 84*
impossible	7	*1. 65*	0	*0*	0	*0*
indeed	15	*3. 55*	2	*0. 53*	7	*1. 65*
X is/was known that/to	6	*1. 42*	0	*0*	1	*0. 23*
more (than)	71 (17)	*16. 81 (4. 02)*	36 (8)	*9. 67 (2. 15)*	6 (4)	*1. 42 (0. 94)*
must	82	*19. 42*	12	*3. 22*	6	*1. 42*
need	10	*2. 36*	2	*0. 53*	2	*0. 47*
obvious	9	*2. 13*	4	*1. 07*	5	*1. 84*

particularly	24	*5. 68*	17	*4. 57*	4	*0. 94*
proved	26	*6. 15*	14	*3. 76*	6	*1. 42*
quite	41	*9. 71*	5	*1. 34*	1	*0. 23*
question	(0) 30	*(0) (7. 10)*	1	*0. 26*	0 (7)	*1. 65*
show / showed	26 / 37	*6. 15 / 8. 76*	8 / 22	*2. 15 / 5. 91*	6 / 9	*1. 42 / 2. 31*
true	9	*2. 13*	2	*0. 53*	3	*0. 71*
will	86	*20. 37*	14	*3. 76*	11	*2. 60*

Table 6. 8. Hedges and boosters with the greatest STTR difference in NLs across time

The greatest difference in booster distribution across the sub-corpora is shown by the modals *must*, *need*, *will*; the epistemic lexical verbs *conclude*, *decided*, *find/found*, *show/showed*, *proved*; the adjectives *certain*, *clear*, *definite*, *impossible*, *obvious*, *true*; the adverbs *actually*, *indeed*, *particularly*, *quite*, the noun *question* and the expressions *X is/was known that*, *more than*, *the fact that*: all these have a higher incidence in the 1901-1939 sub-corpus than in the 1980-2009 corpus, whereas the trend is the opposite for the noun *evidence*, which has a greater frequency in the 1980-2009 sub-corpus than in the 1901-1939 one (my emphasis):

> (1) The detoxicating process acts exclusively in the body *fluids*, in the blood, in the lymphs and in the pericellular lymphatic areas. I must emphasize this especially, because many authors take the view that the diphtheria anti-toxin can also neutralize the poison which has penetrated into the body cells and become established there. (NL2)

> (2) From this it was obvious that, as we had expected, the portions were interchansable-that is to say, presumptive epidermis could become neural plate and presumptive neural plate could become epidermis. From this we can infer not only the very indifferent nature of the cells at this early stage of development; the result allows the much more important conclusion that the transplanted portion must in its new environment be subjected to some kind of influence which determines its subsequent development. (NL7)

> (3) It could be argued that tolerance/response is simply a more refined statement of the homeostasis problem, but language conditions, perceptions and defining the area in this way focuses attention solely towards antigen and away from the more physiological mechanisms and quantitative considerations that are also likely to be enormously important.

(4) This evidence took the form of molecular observations on telomeric DNA not readily explainable by any of the knowledge then about DNA replication or recombination. (NL19)

In academic discourse, hedges and boosters are used to adjust the author's claims by toning down uncertain or potentially risky assertions, emphasising what is thought to be correct, and conveying appropriately collegial attitudes to readers. As we can see in excerpt (3) above, sometimes hedges and boosters can be used simultaneously: in this way, any possible objection is rejected and, at the same time, the audience is left in no doubt. In addition, their use is amplified by the use of the first person pronoun "I", as in (1), or "we" which emphasizes the confidence of the author (s).

Yet what is worth noticing is that, by comparing hedges and boosters across time, it is true that they are used less and less often, but it seems that proportionally, boosters are more frequently used than hedges. This is evidenced in Figure 6. 4. below, where the x-axis indicates the STTR, and the y-axis the sub-corpora. Indeed, while in the 1901-1939 period hedges surpass boosters by 0. 81%, in the 1940-1979 sub-corpus, boosters are used 1.03% more frequently than boosters and in the 1980-2009 sub-corpus boosters exceed hedges by 2. 54%.

Although the NLs seem to become more and more specialized, the fact that boosters are more frequently used than hedges should not come as a surprise: NLs report a type of scientific knowledge that is neither stepping into the academic community for the first time ever, nor needs to be validated, but rather is commonly-shared by the members of the scientific community because of its value, this being the reason why the NL authors have been assigned the Nobel Prize.

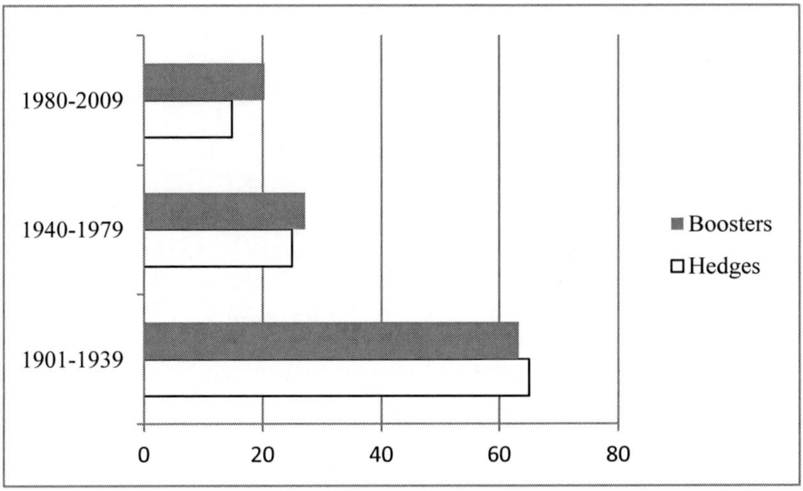

Figure 6. 4. Graphical distribution of hedges and boosters across sub-corpora

5. Conclusions

The main difference between general discourse and specialized discourse is the quantitative presence of specific lexical, syntactic and semantic features (Gotti 2003). Scientific texts present implicit lexical relations which are easily decoded by the scientific community thanks to their background knowledge, whereas popularised texts are constructed in such a way as to explicitly mark those lexical relations, which in specialized texts are implicit (Myers 1991). Yet, when we turn to NLs we are considering a particular form of scientific popularisation targeting a specialized audience stepping outside the professional field of the NL author. This may explain why for these texts there is no need to bridge the gap between specialized and non-specialized registers. Indeed, there seems to be a to-and-fro movement from popularisation to specialization. Firstly, when compared to academic lectures, although both academic lectures and NLs follow a similar genre pattern, NLs seem less instructional than the former. Indeed, NLs are realized in the form of story-telling, where the plot has complicating actions to be resolved by the scientists. This may explain the dramatic rise in frequency of the person pronoun "I" and lexical verbs. These stories presented by the Nobel Laureates seem to be mental "instantations" of the discovery being described, and whose knowledge is nowadays commonly shared by junior and senior members of the medical academic community. Secondly, there is the presence of highly technical elements within NLs which cannot be understood by a layperson. Thirdly, as from the 1940s, there has been a slow but consistent evolution of NLs towards research articles, as they introduce headings, sections, references, i.e. all those formal features characterising scientific academic discourse.

The transformation to the prescribed formal pattern of scientific discourse may reflect a change in the textualized expressions of the authors' stances and opinions expressed by means of hedges and boosters. Boosters and hedges are communicative strategies which can either increase or reduce the force of a statement and indicate the presence of speech-acts, such as doubting and evaluating, rather than merely describing (Salager-Meyer 2006). Hedges are usually employed to reduce the force of an argument, whereas boosters are used to strengthen claims. While hedges allow writers to signal tentativeness in referential information and convey collegial respect for the views of colleagues, boosters express conviction and evidence involvement and solidarity with an audience (Hyland 2004). Generally, NLs show a diminishing use of both hedges and boosters across time. Nevertheless, a closer analysis of such devices in the three sub-corpora in question, divided according to their time-span (that is, 1901-1939, 1940-1979, 1980-2009) and which show a change in the formal structure of NLs, has revealed that the percentage of boosters used in the 1980-2009 period is higher than in the 1901-1939 sub-corpus. The fact that there is a change in the exploitation of boosters in NLs does not necessarily mean that NLS are moving towards a popularised form of the scientific text but rather that NL authors no longer need to mitigate their claims and suppositions because what they are describing is part of commonly-shared knowledge.

Given the type of audience targeted by NLs, the use of such rhetorical devices typical of the popularised forms of scientific discourse such as story-telling and boosters makes it clear that the popularising function of NLs is that of attracting an audience of specialists stepping outside their professional field in order to create networks of teamwork situations (the *invisible college*; cf. Dubois 1985: 72) and thus form networks of teamwork situations (Dubois 1985: 82) with great potentialities as far as future research is concerned.

The results presented here need, of course, to be tried and tested using a more representative corpus and are therefore not conclusive, given the limited coverage and number of texts considered. Yet their implications may invoke new insights regarding the pragmatic effects of popularisation on a professional audience.

BIBLIOGRAPHY

ABCs: National daily newspaper circulation June 2013. *guardian. co. uk*, July 12, 2013.

Algarotti Francesco 1739. *Sir Isaac Newton's Philosophy Explain'd for the Use of the Ladies. In Six Dialogues on Light and Colours*, translated by Elizabeth Carter. London: E. Cave.

Barton, Ruth. 1998. "The Purpose of Science and the Purposes of Popularization in Some English Popular Science Journals of the 1860s". *Annals of Science* 55: 1-33.

Baumlin, J. S. 1994. "Introduction: Positioning ethos in historical and contemporary theory". In J. S. Baumlin and T. F. Baumlin (eds). *Ethos: New essays in rhetorical and critical theory*. Dallas, (TX): Southern Methodist University Press, xi-xxvii.

Baxter, Richard [1665] 1707. "The Reformed Pastor". In *The Practical Works of the Late Reverend and Pious Mr. Richard Baxter*(4 vols.). London.

Beacco, Chantal, Doury, Marianne, Petit, Gerard, and Reboul-Touré, Sandrine. 2002. "Science in media and social discourse: new channels of communication, new linguistic forms". *Discourse Sudies* 4/3: 277-300.

Beeton, Samuel Orchard. 1909. *Beeton's New Dictionary of Every-Day Gardening*. London, Melbourne and Toronto: Ward, Lock and Co. Limited.

Bell, Allan. 1991. *The Language of News Media*. Oxford and Cambridge: Blackwell.

Benwell, Bethan, and Stokoe, Elizabeth. 2006. *Discourse and Identity*. Edinburgh: Edinburgh University Press.

Bhatia, Vijay K. 2008. "Towards critical genre analysis." In Vijay K. Bhatia, John Flowerdew and Rodney H. Jones(eds). *Advances in Discourse Studies*. London and New York: Routledge, 166-177.

—. 2004. *Worlds of Written Discourse. A Genre-based View*. London and New York: Continuum.

Biber, Douglas, Connor, Ulla and Upton, Thomas A. 2007. *Discourse on the Move. Using Corpus Analysis to Describe Discourse Structure.* Amsterdam: John Benjamins Publishing Company.

Biber, Douglas, Johansson, Stig, Leech, Geoffrey, Conrad, Susan and Finegan, Edward. 1999. *Longman Grammar of Spoken and Written English.* Harlow: Longman.

Biber, Douglas, and Finegan, Edward 1989. "Styles of Stance in English: Lexical and Grammatical Marketing of Evidentiality and Affect". *Text* 9: 93-124.

Bowler, Peter J. 2013. *Darwin deleted: imagining a world without Darwin*. Chicago, Illinois: The University of Chicago Press.

—. 2011. *Darwin: Off the record*. London: Watkins.

—. 2009. *Science for all: the popularization of science in early twentieth-century Britain*. Chicago, Ill.: University of Chicago Press.

—. 2005. "Revisiting the eclipse Darwinism". *Journal of the History of Biology* 38, no. 1: 19-32.

—. 1996. *Charles Darwin: the man and his influence*. Cambridge: Cambridge University Press, 1996.

—. 1983. *Eclipse of Darwinism: Anti-Darwinian evolution theories in the decades around 1900*. Baltimore: John Hopkins University Press.

Bright, Henry A. 1891. *A Year in a Lancashire Garden*. London: Macmillan.

Browne, Janet. 2011. "Darwin in caricature: A study in the popularization and dissemination of evolution". *Proceedings of the American Philosophical Society* 145, no. 4: 496-509.

—. 2009. "Looking at Darwin: Portraits and the making of an icon". *Isis* 100, no. 3: 542-570.

Calsamiglia, Helena. 2003. "Editorial. Popularization Discourse". *Discourse Studies* 5/2: 139-146.

Calsamiglia, Helena, and van Dijk, Teun A. 2004. "Popularization discourse and knowledge about the genome". *Discourse Society* 15: 369-389.

Cesiri, Daniela. 2013a. "The Lexicon of Botany Texts in Ireland and England: A Contrastive and Diachronic Case Study from the Late Modern English Period". In McConchie, Roderick W. *et al.* (eds.). *Selected Proceedings of the 2012 Symposium on New Approaches in English Historical Lexis (HEL-LEX 3).* Somerville, MA, Cascadilla Proceedings Project, 35-46.

—. 2013b. "Botany Texts and the Popular Terminology of Plants during the Late Modern English Period in Ireland". In Kermas, Susan and Thomas Christiansen (Eds.). *The Popularization of Specialized Discourse and Knowledge Across Communities and Cultures*. Bari: Edipuglia, 67-84.

—. 2012a. "Herbals, Gardening Books and Botanic Terminology in Ireland: a Preliminary Description for a Linguistic Investigation". In Chiavetta, Eleonora and Silvana Sciarrino. (eds.). *The Popularisation of Botanical, Legal and Commercial Language*. Roma: Edizioni XL, 13–24.

—. 2012b. "Investigating the Development of ESP through Historical Corpora: the Case of Archaeology Articles Written in English during the Late Modern Period (and Beyond?)". In Tyrkkö, Jukka *et al.* (eds.). *Proceedings of the Helsinki Corpus Festival*. (Studies in Variation, Contacts and Change in English 10). Helsinki: Research Unit for Variation, Contacts, and Change in English. Accessed June 2013. http://www. helsinki. fi/varieng/journal/volumes/10/cesiri/.

Cherry, Roger D. 1998. "Ethos versus Persona: Self-representation in written discourse".*Written Communication* 15: 384-410.

Chiavetta, Eleonora. 2013. "'Gardening for the ignorant': Mrs C. W. Earle and the Popularization of Gardening Matters". In Kermas, Susan, and Thomas Christiansen (eds). *The Popularization of Specialized Discourse and Knowledge Across Communities and Cultures*. Bari: Edipuglia, 55-66.

—. 2012. "Jane Webb Loudon Populariser of the Language of Botany". In Chiavetta, Eleonora and Silvana Sciarrino (eds). *The Popularization of Botanical, Legal and Commercial Language.* Roma: XL Edizioni, 25-40.

—. 2008. "'A patient gleaner in the fields of knowledge': a discourse analysis of two late 19th century gardening books". In Kermas, Susan, and Maurizio Gotti (eds). *Socially-Conditioned Language Change: Diachronic and Synchronic Insights*. Lecce: Edizioni del Grifo, 137-155.

Chichester Hart, Henry. 1898. *Flora of the County Donegal or List of the Flowering Plants and Ferns with their Localities and Distribution.* Dublin: Sealy, Bryers and Walker.

Christiansen, Thomas. 2013. "Cohesive conjunctions and their function in the discourse of the popularization of science: Charles Darwin's correspondence on evolution and related matters" in Kermas, Susan and Thomas Christiansen (eds.). *The Popularization of Scientific Discourse in a Changing World*. Bari: Edipuglia, 131-145.

—. 2012. "The metaphoric properties of the term 'blood' in the works of Charles Darwin and Edith Durham: A comparison of conceptual frames" in Chiavetta, Eleonora and Silvana Sciarrino (eds.). *The Popularization of Botanical, Legal and Commercial Language*. Rome: XL Edizioni, 41-62.

—. 2010. "Letter writing as an instrument of scientific discourse: the correspondence of Charles Darwin". In Cottone Margherita and Eleonora Chiavetta (eds) *La Scrittura Epistolare in Europa dal Medievo ai nostri giorni: Generi, modelli e trasformazioni*. Acireale: Bonnano Editore, 235-261.

—. 2009. "The Creation of Evolution: Religion and Science in the Lexis and Conceptual Frames of the Works of Charles Darwin". In Gotti Maurizio (ed.). *Trading Identities. Commonality and Individuality in English Academic Discourse*. Bern: Peter Lang, 255-278.

Ciapuscio, Guiomar E. 2003. "Formulation and Reformulation Procedures in Verbal Interactions between Experts and (Semi)-laypersons". *Discourse Studies* 5: 207-233.

Clark, Timothy. 1980. "Mrs C. W. Earle, (1836-1925), a reappraisal of her work". *Garden History. The Journal of the Garden History Society.* VIII: 75-83.

Coffey, Jane, and Atkinson, Paul A. 1996. *Making sense of qualitative data: Complimentary research strategies.* Thousand Oaks, CA: Sage.

Cohen, Stanley, Levi-Montalcini, Rita and Hamburger, Viktor. 1954. "A Nerve Growth-Stimulating Factor Isolated from Sarcom AS 37 and 180". *Proceedings of the National Academy of Sciences* 40/10: 1014-1018.

Constantine, Stephen. 1981. "Amateur Gardening and Popular Recreation in the 19th and 20th Centuries". *Journal of Social History* 14 (3): 387-406.

Cooter, Roger, and Pumfrey, Stephen. 1994. "Separate Spheres and Public Places: Reflections on the History of Science Popularisation and Science in Popular Culture". *History of Science*32/3: 237-267.

Cope Jackson I. 1954. "Joseph Glanvill, Anglican apologist: Old ideas and new style in the Restoration". *PMLA* vol. 69 no. 1: 223-250.

Davidson, James. 2012. "Create GB; Dominic Davies: mastering the art of still life photography". *weheart. co. uk*, July 25.

Darwin, Charles R. 1881. *The Formation of Vegetable Mould through the Action of Worms*, with Observations on their Habits. London: John Murray.

—. Letter to T. H. Huxley from Down. December 7, 1862. Available on the Darwin Correspondence Project, http: //www. darwinproject. ac. uk/entry-3848. (Accessed May 30, 2013).

—. [1859] 1996. *The origin of species*. Oxford: Oxford University Press.

—. Letter from C. R. Darwin to sister Catherine. July 24, 1842. *English Heritage MS Darwin Collection*, 4: 1.

De Fontenelle, Bernard le Bovier.1688, *Conversations on the Plurality of Worlds*, translated by Aphra Behn, London.

de Oliveira, Janaina M., and Pagano, Adriana S. 2006. "The research article and the science popularization article: a probabilistic functional grammar perspective on direct discourse representation". *Discourse Studies* 8: 627-646.

Deroey, Katrien L. B., and Taverniers, Miriam. 2011. "A corpus-based study of lecture functions". *Modernaspråk*2: 1-22.

Di Martino, Gabriella. 1998. "The 'New Science' and the new language in Seventeenth century England". In Sornicola, Rosanna and Dieter Stein (eds). *Essays in Honour of Thomas Frank*. Amsterdam: Benjamins, 129-149.

Dickie, George. 1864. *A Flora of Ulster and Botanist's Guide to the North of Ireland*. Belfast: Aitchinson.

Dobzhansky, Theodosius. 1937. *Genetics and the Origin of Species*. New York: Columbia University Press.

Dolan, Terence P. (ed.) 1998. *A Dictionary of Hiberno-English*. Dublin: Gill & Macmillan.

Dörnyei, Zoltán. 2007. *Research Methods in Applied Linguistics: Quantitative, Qualitative, and Mixed Methodologies*. Oxford: Oxford University Press.

Dossena, Marina. 2008. "«The times they're a-changing»: The Abolition of Feudal Tenure (Scotland) Act 2000 and Linguistic Strategies of Popularisation". In Bhatia, Vijay K., Christopher N., Candlin, and Paola, Evangelisti Allori (eds). *Language, Culture and the Law*. Bern: Peter Lang, 187-206.

Dubois, Beth L. 1985. "Popularization at the Highest Level: Poster Sessions at Biomedical Meetings". *International Journal of the Sociology of Language* 56: 67-84.

Earle, C. W. 1903. *A Third-Pot Pourri*. London: Smith Elder.

—. 1899. *More-Pot Pourri*. London: Smith Elder.

—. 1897. *Pot-Pourri from a Surrey Garden*. London: Smith Elder.

Ellegård, Alvar. 1958. *Darwin and the general reader. The reception of Darwin's theory of evolution in the British periodical press, 1859-1872*. Göteborg: Elanders Boktryckeri Aktiebolag.

Elliston, Allen D. and Hatfield, Gabrielle. 2004. *Medicinal Plants in Folk Tradition: An Ethnobotany of Britain and Ireland*. Portland, Oregon: Timber Press.

Fairclough, Norman and Chouliaraki, Lilie. 1999. *Discourse in late modernity: Rethinking critical discourse analysis*. Edinburgh: Edinburgh University Press.

Fairclough, Norman. 2006. *Language and Globalization*. London: Routledge.

—. 2003. *Analysing Discourse: Textual Analysis for Social Research*. London: Routledge.

—. 1995a. *Media Discourse*. London: Edward Arnold.

—. 1995b. *Critical Discourse Analysis*. Boston: Addison Wesley.

—. 1992. *Discourse and Social Change*, Cambridge: Polity Press.

Fisher, Ronald A. 1930. *The Genetical Theory of Natural Selection*. Oxford: Clarendon Press.

Flowerdew, John. 1994. "Research of relevance to second language lecture comprehension-an overview". In Flowerdew, John (ed.). *Academic listening: research perspectives*. Cambridge: Cambridge University Press, 7-29.

Fraser, Mitchell William. 1932. *English Pulpit Oratory from Andrewes to Tillotson: A Study of the Literary Aspects*. London, Society for promoting Christian Knowledge; New York and Toronto: The Macmillan Co.

Gezi, Lourdes Jahrling. 1959. *Changes in Glanvill's Prose*, a Thesis submitted to the Dept of English of Stanford University.

Giannoni, Davide S. 2008. "Popularizing features in English journal editorials". *English for Specific Purposes*27/2: 212-232.

Giunchi, Paola. 2002. "Information or Misinformation? 'Translating' Medical Research Papers into Web-posted Accounts". In Cortese, Giuseppina, and Philip Riley (eds). *Domain-specific English*. Bern: Peter Lang, 271-293.

Glanvill, Joseph. [1661] 1931. *The Vanity of Dogmatizing*, reproduced from the Edition of 1661, with a bibliographical note by Moody E. Prior, New York, published for the Facsimile Text Society by Columbia University Press.

—. 1678. *Essay concerning preaching: written for the direction of a young divine*. By Jos. Glanvill chaplain to King Charles II and fellow of the Royal Society. The second edition, London, printed for C. Brome at the Gun, at the West-end of St. Paul's Church, 1703.

—. 1676. *Essays on Several Important Subjects in Philosophy and Religion*, London, printed by J. D. for J. Baker.

—. 1676a. *Seasonable reflections and discourses in order to the conviction, and cure of the scoffing, and infidelity of a degenerate age.* By Jos. Glanvill, chaplain in ordinary to his Majesty. London, printed by R. W. for H. Mortlock at the Phoenix in St. Paul's Churchyard, at the White-Hart in Westminster-Hall.

—. [1668] 1958. *Plus Ultra or the Progress and Advancement of Knowledge Since the Days of Aristotle*, a facsimile reproduction with an introduction by Jackson I. Cope, Gainsville, Florida, Scholar's Facsmiles and Reprints.

—. 1665. *Scepsis Scientifica or Confest Ignorance, the Way to Science,* printed by E. Cotes for Henry Eversden.

Glaser, Rosemarie. 1995. *Linguistic Features and Genre Profiles of Scientific English*. Berlin, Bern, New York, Paris, Wien: Peter Lang.

Glick, Thomas F.2010. *"What about Darwin?": All Species of Opinion from Scientists, Sages, Friends, and Enemies Who Met, Read, and Discussed the Naturalist Who Changed the World.* Baltimore (MD): The Johns Hopkins University Press.

Goodfellow, Sarah. 1996. "'Such Masculine Strokes': Aphra Behn as Translator of *A Discovery of New Worlds*". *Albion, A Quarterly Journal concerned with British Studies*, 28 n. 2: 229-250.

Gotti, Maurizio. 2013. "The analysis of popularization discourse: Conceptual changes and methodological evolutions." In Kermas, Susan and Thomas Christiansen (eds). *The Popularization of specialized discourse and knowledge across communities and cultures*. Bari: Edipuglia, 9-32.

—. 2003. *Specialized Discourse: Linguistic Features and Changing Conventions*. Bern: Peter Lang.

—. 1996. *Robert Boyle and the Language of Science*. Milano: Guerini Scientifica.

Greenslet, Ferris. 1900. *Joseph Glanvill: a Study in English Thought and Letters of the Seventeenth Century.* New York: The Columbia University Press.

Grego, Kim. 2013. "'The physics you buy in supermarkets'. Writing science for the general public: the case of Stephen Hawking'." In Kermas, Susan and Thomas Christiansen (eds). *The Popularization of specialized discourse and knowledge across communities and cultures*. Bari: Edipuglia, 149-172.

Haldane, John Burdon S. 1932. *The Causes of Evolution*. London: Longmans Green & Co.

Halliday, Michael Alexander K. 2004. *An Introduction to Functional Grammar* (3rd edition revised by Christian M. I. M. Matthiessen). London: Edward Arnold.

Harbottle, Bryan. 2004. *The bugs in Darwin's bed*. London: Athena.

Harvey, John H. 1985. "The First English Garden Book. Mayster Jon Gardener's Treatise and Its Background". *Garden History.* 13: 83-101.

Hilgartner, Stephen. 1990. "The Dominant View of Popularization: Conceptual Problems, Political Uses". *Social Studies of Science*20/3: 519-539.

Horenstein, Sidney. 2009. "The 'popular press' responds to Charles Darwin, The Origin of Species and his other works". *Evolution: Education and Outreach* 2, no. 1: 107-116.

Hoyles, Martin. 1994. *Gardeners Delight. Gardening Books from 1560 to 1960*. London-Bouler-Colorado: Pluto Press.

Horwood, Catherine. 2010. *Gardening Women. Their Stories from 1600 to the Present*. London: Virago.

Hughes, Geoffrey. 1988. *Words in Time*. Oxford: Blackwell.

Hull, David L., Tessner Peter D. and Diamond, Arthur M. 1978. "Planck's Principle". *Science*, New Series, 202, no. 4369: 717-723.

Hull, David L. 1973. *Darwin and his critics: The reception of Darwin's theory of evolution by the scientific community*. Cambridge, Mass.: Harvard University Press.

Hunston, Susan. 2002. *Corpora in Applied Linguistics*. Cambridge: Cambridge University Press.

Hunston, S. and Thompson, G. (eds.). 2000. *Evaluation in Text: Authorial Stance and the Construction of Discourse*. Oxford: Oxford University Press.

Hunt, Morgan, Thomas. 1916. *A Critique of the Theory of Evolution*. Princeton (NJ): Princeton University Press.

Hunt Morgan, Thomas, Sturtevant Alfred H., Mulle, Hermann J. and Bridges, Calvin B. 1915. *The Mechanism of Mendelian Heredity*. New York: Henry Holt.

Huxley, Julian. 1942. *Evolution: the modern synthesis*. London: Allen and Unwin.

Huxley, Thomas H. 1893. *Darwiniana. Essays by Thomas H. Huxley*. London: Macmillan & co.

—. 1863. *Evidence as to man's place in nature*. London: Williams and Norgate.

—. 1859. "Darwin on the origin of species". *The Times*, December 26: 8-9.

Hyland, Ken. 2004. *Disciplinary Discourses: Social Interactions in Academic Writing*. Ann Arbor: University of Michigan Press.

—. 2001. "Humble Servants of the discipline? Self-mention in research articles". *English for Specific Purposes* 20: 207-226.

—. 2000b. "Hedges, Boosters and Lexical Invisibility: Noticing Modifiers in Academic Texts". *Language Awareness* 9/4: 179-197.

IFABC, The International Federation of Audit Bureaux of Circulations. Circulation Data For National Newspaper. January 17, 2013. Available on www.ifabc.org/site/assets/media/National-Newspapers_total-circulation_IFABC_17-01-13. xls. (Accessed May 30, 2013).

Jacobi, Daniel. 1990. "Les series superordonnées dans les discours de vulgarisation scientifique."*Langages*98: 103-114.

Johnson, Phillip E. 1993. *Darwin on Trial*. Nottingham: InterVarsity Press.

Jones, Richard Foster. 1951. *The Seventeenth Century: Studies in the History of English Thought and Literature from Bacon to Pope*. Stanford, California: The Stanford University Press.

K'Eogh, John. 1735. *Botanologia Universalis Hibernica*. Cork: George Harrison.

Keeble, N. H. 1987, *The Literary Culture of Nonconformity in Later Seventeenth-century England*. Leicester: Leicester University Press.

Kelly, Alfred. 1981. *The descent of Darwin: the popularization of Darwinism in Germany 1860-1914*. Chapel Hill, N. C.: University of North Carolina Press.

Kermas, Susan. 2006. "From Herbal to Advert". In Brownlees, Nicholas (ed). *News Discourse in Early Modern Britain*. Bern: Peter Lang, 175-197.

Kiewra, Kenneth. A. 2002. "How classroom teachers can help students learn and teach them how to learn". *Theory into Practice* 41: 71-80.

Kim Jong Il, Brief History. 1998. Pyongyang: Foreign Languages Publishing House. Available on http: //www. korea-dpr. com/lib/103. pdf.

Knudsen, Susanne. 2003. "Scientific metaphors going public". *Journal of Pragmatics* 35: 1247–1263.

Kockelman, Paul. 2004. "Stance and Subjectivity". *Journal of Linguistic Anthropology* 14, Issue 2: 127-150.

Kress, Gunther and Hodge, Robert. 1979. *Language as ideology*. London: Routledge and Kegan Paul.

Krogh, August. 1922. *The anatomy and physiology of capillaries*. London: Oxford University Press.

Labov, William. 2002. "Ordinary events". At: http://www.ling.upenn. edu/~wlabov/Papers/OE. pdf.

—. 2001. "Uncovering the event structure of narrative". At: http://www. ling. upenn. edu/~wlabov/uesn. pdf.

Leake, Erik. 2012. “Science as Sound Bites: The Lancet Iraq Casualty Reports and Prefigured Accommodation”. *Technical Communication Quarterly*21/2: 129-144.

Leane, Elizabeth. 2001. “Knowing quanta: the ambiguous metaphors of popular physics”. *The Review of English Studies*52/207: 411-431.

Lee, Joseph. J. 2009. “Size matters: an exploratory comparison of small-and large-class university lecture introductions”. *English for Specific Purposes* 28: 42-57.

Lightman, Bernard V. 2010. “Darwin and the popularization of evolution”. *Notes & Records of the Royal Society* 64, no. 1: 5-24.

—. 2009. *Evolutionary naturalism in Victorian Britain: The ‘Darwinians’ and their critics*. Burlington, Vermont, USA; Farnham, Surrey, England: Ashgate.

—. 2007. *Victorian popularizers of science: Designing nature for new audiences*. Chicago: University of Chicago Press.

Lightman, Bernard V. and Shteir, Ann. 2006. *Figuring it out: science, gender and visual culture*. Hanover, New Hampshire: Dartmouth College Press; Hanover and London: University Press of New England.

Mackay, James T. 1836. *Flora Hibernica*. Dublin: Curry Jun and Co.

MacLeod, Dawn. 1982. *Down-to-Earth Women*. Edinburgh: William Blackwood.

Masiola, Rosanna and Tomei, Riccardo. 2009. *West of Eden: Botanical Discourse, Contact Languages and Translation*. Newcastle upon Tyne: Cambridge Scholars Publishing.

Mazzotti Massimo. 2004, “Newton for ladies: gentility, gender and radical culture” in *BJHS* 37 (2): 119-146.

Miles, Matthew B. and Huberman, A. Michael. 1994. *Qualitative data analysis: An expanded sourcebook.* Thousand Oaks: Sage.

Moirand, Sophie. 2003. “Communicative and Cognitive Dimensions of Discourse on Science in the French Mass Media”. *Discourse Studies* 5: 175-206.

Moore, James. 2002. “Charles Darwin”. In Ferngren, Gary B. (ed.) *Science and Religion: A Historical Introduction*. Baltimore: (MD), John Hopkins University Press, 208-218.

Mordavsky Caleb, Amanda. (ed.). 2007. *(Re) Creating Science in Nineteenth-Century Britain.* Newcastle: Cambridge Scholars Publishing.

Myers, Greg. 2003. “Discourse studies of scientific popularization: questioning the boundaries”. *Discourse Studies*3: 265-279.

—. 1991. “Lexical Cohesion and Specialized Knowledge in Science and Popular Science Texts”. *Discourse Processes*14: 1-26.

Numbers, Ronald L. and Stenhouse, John (eds). 1999. *Disseminating Darwinism: the role of place, race, religion, and gender*. Cambridge: Cambridge University Press.

The Oxford English Dictionary. Online edition. Accessed June 2013. http://www. oed. com.

Parkinson, Jean, and Adendorff, Ralph. 2004. "The use of popular science articles in teaching scientific literacy". *English for Specific Purposes* 23: 379-396.

Paul, Danette. 2004. "Spreading Chaos: The Role of Popularizations in the Diffusion of Scientific Ideas". *Written Communication* 21: 32-68.

Peacham, Henry. [1593] 1954, *The Garden of Eloquence* (a facsimile reproduction with an introduction by William G. Crane, Scholars' facsimiles and Reprints.

Peachey, Kevin. "Sir Winston Churchill to feature on new banknote". *bbc. co. uk*, April 26, 2013.

Power, Thomas. *et al*. 1845. *Contributions Towards a Fauna and Flora of the County of Cork*. Cork: Purcell.

Praeger, Robert L. 1934. *The Botanist in Ireland*. Dublin: Hodges, Figgis & Co.

Pramling, Niklas and Säljö, Roger. 2007. "Scientific Knowledge, Popularisation, and the Use of Metaphors: Modern genetics in popular science magazines". *Scandinavian Journal of Educational Research* 51/3: 275-295.

Raichvarg, Daniel and Jacques, Jean. (1991). *Savants et ignorants-une histoire de la vulgarization des sciences*. París: Editions du Seuil.

Rankin, Jennifer. "Jane Austen could grace £10 banknotes, Mervyn King says". *guardian. co. uk*, June 25, 2013.

Rayson, Paul. 2009. "Wmatrix: a web-based corpus processing environment". At http: //ucrel. lancs. ac. uk/wmatrix/.

Redgrove, Herbert. S. and Redgrove, I. M. L. 1921. *Joseph Glanvill and psychical research in the seventeenth century*. London: W. Rider and Son.

Rhode, Eleanour Sinclair. 1922. *The Old English Herbals*. London: Longman, Green and Co.

Sackville-West, Vita. 1939. *Country Notes*. London: Michael Joseph Ltd.

Salager-Mayer, Françoise. 2006. *I Think that Perhaps you Should: A Study of Hedges in Written Scientific Discourse*. At www.exchanges.state. gov/education/engteaching/pubs/BR/functionalsec3_8. htm.

Scott, Mike. 2008. *WordSmith Tools version 5*. Liverpool: Lexical Analysis Software.

Seaton, Beverly. 1979. "The garden autobiography". In *Garden History – The Journal of the Garden History Society*, VII, n. 1: 101-114.

Sharma, P. Gopal. 1972. "Popularization of science and communication crisis in languages". *Babel* 18/3: 8-10.

Skorcynska, Hanna. 2001. "Metaphor in scientific business journals and business periodicals: an example of the scientific discourse popularisation". *Iberica* 3: 43-60.

Smith, Jonathan. 2006. *Charles Darwin and Victorian visual culture*. Cambridge: Cambridge University Press.

Spencer, Herbert. 1864. *The Principles of Biology*, Vol I. London: Williams and Norgate.

Sperber, Dand and Wilson, Deirdre. 1986. *Relevance, Communication and Cognition*. Oxford: Blackwell.

Sprat, Thomas. 1667. *The History of the Royal Society of London, for the Improving of Natural Knowledge*, London, printed by T. R. for J. Martyn.

Stewart, Samuel A. and Corry, Thomas H. 1888. *A Flora of the North-East of Ireland*. Cambridge: MacMillan & Bowes.

Stuessy, Tod F. 2009. *Plant Taxonomy: The Systematic Evaluation of Comparative Data*. New York: Columbia University Press.

Synnott, Donal. 1997. "Botany in Ireland". In Foster, John Wilson (ed.), *Nature in Ireland: A Scientific and Cultural History*. Dublin: The Lilliput Press, 157-183.

Teichmann, Jurgen et al. (eds). 2004. *From the itinerant lectures of the 18th century to popularizing physics in the 21st century – exploring the relationship between learning and entertainment,* Proceedings of a conference held in Pognana sul Lario, Italy, June 1-6 2003. Munich, Winnipeg, Oldenburg.

The Shorter Oxford English Dictionary on Historical Principles, prepared by William Little, H. W. Fowler and Jessie Coulson, revised and edited by C. T. Onions. 1973. Vol. II. Oxford: Clarendon Press.

*The Times Digital Archive (1785-2006).*London: Natkins.

Thoiron, Philippe, and Béjoint, Henri. 1991. "La place des reformulations dans les texts scientifiques". *Meta: journal des traducteurs/Meta: Translators' Journal* 36/1: 101-112.

Uglow, Jenny. 2005. *A Little History of British Gardening*. London: Pimlico.

van Dijk, Teun A. 2008. *Discourse and Power*. Houndsmills: Palgrave.

—. 1988. *News as Discourse*, Hillsdale, NJ: Erlbaum.

Vicentini, Alessandra, Grego, Kim and Russo, Daniele. 2013. "The Pro. Bio. Dic. (*Prototype of a Bioethics Dictionary*) project: Building a corpus of popular and specialized bioethics texts."*JAHR-Proceedings from the International Conference* The Language of Medicine – from Its Genesis to the Culture and Ethics of Communication 4, no. 7: 339-354.

Webb, David A. 1986. "The Hey-Day of Irish Botany, 1866-1916". *The Scottish Naturalist*: 123-134.

Willey, Basil. 1952. *The Seventeenth Century Background: Studies in the Thought of the Age in Relation to Poetry and Religion*. New York: The Columbia University Press.

Wilkins, John. 1653. *Ecclesiastes, or a Discourse concerning the Gift of Preaching*, the fourth edition, London.

Wodak, Ruth (ed.) 2013. *Critical Discourse Analysis*. 4 vol. London: Sage.

Wright, Joseph. (ed.). 1898-1905. *The English Dialect Dictionary*. 6 vols. London: Henry Frowde.

CONTRIBUTORS

Daniela Cesiri holds a PhD in English Language and Linguistics. She is currently Researcher and Lecturer at the University “Ca’ Foscari” of Venice (Italy). Her main research interests focus on the use of corpus linguistics and discourse analysis to the diachronic and synchronicstudy of English for Specific and for Academic Purposes. Her most recent publications include her first monograph, entitled *Nineteenth-Century Irish English. A Corpus-Based Linguistic and Discursive Analysis*, (Lampeter, Wales & New York: The Edwin Mellen Press, 2012) as well as two articles related to the topic presented in this volume, namely “Botany texts and the popular terminology of plants during the Late Modern English period in Ireland” (in S. Kermas and T. Christiansen, *The Popularization of Specialized Discourse and Knowledge across Communities and Cultures*, Bari: Edipuglia, 2013) and “The Lexicon of Botany Texts in Ireland and England: A Contrastive and Diachronic Case Study from the Late Modern English Period” (in McConchie Roderick W. et al., *Selected Proceedings of the 2012 Symposium on New Approaches in English Historical Lexis (HEL-LEX 3),* Somerville, MA: Cascadilla Proceedings Project).

The research of **Eleonora Chiavetta**, Associate Professor (now retired) of English Language at the University of Palermo, focuses on specialized discourse, genre, letter writing, and discourse analysis in a diachronic perspective. She has written many essays on these topics. She is particularly interested in botanical and gardening texts written by women in the 19th century and in the early 20th century. Among her latest publications, she has co-edited, together with Margherita Cottone, the collection of essays on *La scrittura epistolare in Europa* (Acireale-Roma, 2010), and, with Silvana Sciarrino, the collection of essays, *The Popularization of Botanical, Legal, and Commercial Language* (Roma, 2012).

Since 1987, **Thomas Christiansen** has taught in various positions at various institutions in Apulia, Italy, and in Poland. In 2001, he completed his PhD at Salford (UK). He has published books and a number of articles on various areas of linguistics including systemic linguistics and functional grammar, varieties of English, ELF, teaching English, language testing, and analysis of different corpora, including spoken discourse. Since 2006, he has been a full-time researcher and adjunct professor in English Language and Translation at the Università del Salento (Lecce, Italy). He has given papers at many national and international countries including: Albania, Austria, Denmark, Finland, Turkey, and the UK.

Gabriella Di Martino is Full Professor of English Language and Linguistics at the Department of Political Science, Naples University, Federico II. She is Senior member of the teaching staff and scientific coordinator of the PhD course in English for Special Purposes and also Local Project Coordinator of National Interest Research Programmes. Her main current research interests cover the evolution of English in non-literary prose as well as synchronic and diachronic discourse analysis covering general linguistics issues in ESP.

Kim Grego is an Assistant Professor of English Linguistics and Translation at the University of Milan, where she teaches English. Her interests include translation studies, ESP, discourse analysis and genre analysis. Recent publications are: *Specialized Translation: Theoretical issues, operational perspectives*, (Polimetrica, Monza , 2010), "'The physics you buy in supermarkets': writing science for the general public: the case of Stephen Hawkings", in S. Kermas and T. Christiansen (eds), *The popularization of specialized discourse and knowledge across communities and cultures* (Bari, Edipuglia, 2013) and "Intercultural and Ideological Issues in Lexicography: A Prototype of a Bioethics Dictionary" in Facchinetti, R. (ed.), *Cultural identities in English lexicography* (Cambridge Scholars Publishers, Newcastle, 2012).

Stefania M. Maci is Aggregate Professor of English Language and Translation at the University of Bergamo, where she teaches English linguistic courses at graduate and undergraduate level. She is member of CERLIS (Research Centre on Specialized Languages, coordinated by Prof. Maurizio Gotti), CLAVIER (The Corpus and Language Variation in English Research Group), BAAL (British Association of Applied Linguistics), and AIA (Associazione Italiana di Anglistica). Her research is focussed on the study of the English language in academic and professional contexts, with particular regard to the analysis of medical discourse, legal discourse and tourism discourse. Her most recent publications include: "Glocal Features of In-flight magazines" (2012), "Arbitration in action: the display of arbitrators' neutrality in witness hearings" (2012); "The Discussion Section of Medical Research Articles: A Cross Cultural Perspective" (2012); "Fast-Track Publications: The Genre of Medical Research letters" (2012); "The Genre of Medical Conference Posters" (2012); "Poster Makers Should Think as Much about Show Business as Science. The Case of Medical Posters in a Diachronic Perspective" (2012); and the monographs, *Tourism Discourse: professional, promotional, digital voices*" (2013); *The Language of Tourism* (2010), and *The Linguistic Design of Mary Magdalene* (2008).